BI 3246655 2

D1610840

OXFORD GEOGRAPHICAL AND ENVIRONMENTAL STUDIES

Editors: Gordon Clark, Andrew Goudie, and Ceri Peach

CONFLICT, CONSENSUS, AND RATIONALITY IN ENVIRONMENTAL PLANNING

ALSO PUBLISHED BY
OXFORD UNIVERSITY PRESS
IN THE OXFORD GEOGRAPHICAL AND
ENVIRONMENTAL STUDIES SERIES

Island Epidemics
Andrew Cliff, Peter Haggett, and Matthew Smallman-Raynor

Pension Fund Capitalism
Gordon L. Clark

Class, Ethnicity, and Community in Southern Mexico
Oaxaca's Peasantries
Colin Clarke

Indigenous Land Management in West Africa
An Environmental Balancing Act
Kathleen Baker

Cultivated Landscapes of Native
North America
William E. Doolittle
Paperback

Cultivated Landscapes of Native Amazonia
and the Andes
William M. Denevan

Globalization and Integrated Area
Development in European Cities
Frank Moulaert
Paperback

Cultivated Landscapes of Middle America
on the Eve of Conquest
Thomas M. Whitmore and B.L. Turner II

Globalization and Urban Change
Capital, Culture, and Pacific Rim Mega-Projects
Kris Olds
Paperback

A Critical Geography of Britain's State Forests
Judith Tsouvalis

The Modalities of European Union Governance
New Institutional Explanations of
Agri-Environmental Policy
Alun Jones and Julian Clark

Conflict, Consensus, and Rationality in Environmental Planning

An Institutional Discourse Approach

Yvonne Rydin

OXFORD
UNIVERSITY PRESS

OXFORD
UNIVERSITY PRESS

Great Clarendon Street, Oxford OX2 6DP

Oxford University Press is a department of the University of Oxford.
It furthers the University's objective of excellence in research, scholarship,
and education by publishing worldwide in

Oxford New York

Auckland Bangkok Buenos Aires Cape Town Chennai
Dar es Salaam Delhi Hong Kong Istanbul Karachi Kolkata
Kuala Lumpur Madrid Melbourne Mexico City Mumbai Nairobi
São Paulo Shanghai Taipei Tokyo Toronto

Oxford is a registered trade mark of Oxford University Press
in the UK and certain other countries

Published in the United States
by Oxford University Press Inc., New York

First published 2003

British Library Cataloguing in Publication Data
Data available

Library of Congress Cataloging in Publication Data
Rydin, Yvonne, 1957–
Conflict, consensus, and rationality in environmental planning : an insitutional
discourse approach / Yvonne Rydin.
p. cm.–(Oxford geographical and environmental studies)
Includes bibliographical references and index.
1. Envronmental policy–Social aspects. 2. Environental
protection–Planning–Social aspects. I. Title. II. Series.
GE170 .R93 2003 363.7′05–dc21 2002033670
ISBN 0-19-925519-9

1 3 5 7 9 10 8 6 4 2

Typeset by SNP Best-set Typesetter Ltd., Hong Kong
Printed in Great Britain
on acid-free paper by
Biddles Ltd, Guildford and King's Lynn

EDITORS' PREFACE

Geography and environmental studies are two closely related and burgeoning fields of academic enquiry. Both have grown rapidly over the past few decades. At once catholic in its approach and yet strongly committed to a comprehensive understanding of the world, geography has focused upon the interaction between global and local phenomena. Environmental studies, on the other hand, have shared with the discipline of geography an engagement with different disciplines, addressing wide-ranging and significant environmental issues in the scientific community and the policy community. From the analysis of climate change and physical environmental processes to the cultural dislocations of post-modernism across the landscape, these two fields of enquiry have been at the forefront of attempts to comprehend transformations taking place in the world, manifesting themselves at a variety of interrelated spatial scales.

The Oxford Geographical and Environmental Studies series aims to reflect this diversity and engagement. Our goal is to publish the best original research in the two related fields and, in doing so, demonstrate the significance of geographical and environmental perspectives for understanding the contemporary world. As a consequence, our scope is deliberately international and ranges widely in terms of topics, approaches, and methodologies. Authors are welcomed from all corners of the globe. We hope the series will assist in redefining the frontiers of knowledge and build bridges within the fields of geography and environmental studies. We hope also that it will cement links with issues and approaches that have originated outside the strict confines of these disciplines. In so doing, our publications contribute to the frontiers of research and knowledge while representing the fruits of particular and diverse scholarly traditions.

Gordon L. Clark
Andrew Goudie
Ceri Peach

ACKNOWLEDGEMENTS

This book has been over a decade in the making, if not in the writing. During that time I have benefited from working with many people, whom I would like to thank here. First and foremost, I wish to acknowledge my intellectual debt to George Myerson, who opened up the whole world of discourse to me. From the origins of this particular dialogue in early joint articles, through our book *The Language of Environment*, to his generous reading of the whole typescript, George has been a central part of my engagement with these ideas. Secondly, I would like to thank Jacquie Burgess for being the most helpful of publisher's referees, not only by reading the typescript twice but by providing detailed and summary comments. Her contribution has undoubtedly improved the book; my lack of understanding of or response to her comments may explain some of its failings. I have over the last decade had the privilege to work with many colleagues on a number of research projects, which has fed into the development of the ideas expressed here. I would like to thank all those involved in the EU project, 'The Economic and Cultural Conditions of Decision Making for the Sustainable City', the EU project, 'Promoting Action for Sustainability Through Indicators at the Local Level in Europe' (Pastille), and the ESRC project, 'The Metropolitan Governance and Community Study'. I particularly acknowledge the discussions I had with Mark Pennington as part of a Leverhulme project on local environmental decision-making on the potential and limitations of the rational choice framework. More generally, a number of people have contributed to the broader context for this work. Patsy Healey has played a pivotal role in the planning research community, generating an atmosphere of intellectual excitement and enquiry. I have benefited from many conferences and seminars organized under her auspices. The LSE continues to be the best of places to write and research. I wish to thank my colleagues in the Department of Geography and Environment, as well as the many students—both M.Sc. and research students—who generate such interesting discussions. At Oxford University Press I would like to thank Anne Ashby for all her editorial help in steering the book through its various stages. And finally, all thanks and love to George, Simon, and Eleanor for the best context of all.

CONTENTS

LIST OF FIGURES

1

Environmental Planning: Introducing the Discourse Approach

A popular adage of the late nineteenth century in the developed world was that 'where there's muck, there's brass'. Economic activity and wealth creation was associated with dirt and pollution. This was true of the factories, mines, and quarries as well as the cities where growing numbers of people lived. Human waste was another aspect of this 'muck'. And with human waste came disease and public health hazards. The response to this was the first era of environmental planning, focused on regulating industrial pollution and managing residential environments to remove human waste through sewerage systems and improve the circulation of fresh air. While a decline in public disease episodes provided some evidence of the success of such planning, problems of ill health among urban residents and environmental pollution persisted into the twentieth century. The growth of production and consumption during this century generated new problems, with rapidly escalating amounts of increasingly toxic manufactured waste to be disposed of, much of it extremely toxic. Energy consumption also increased significantly.

By the end of the twentieth century a new term was being coined to describe the goal for a new era of environmental planning: sustainable development. This captured the idea that environmental systems were suffering irreversible and possibly catastrophic damage due to the burdens of our contemporary patterns of economic and social organization. Particular emphasis was placed on the depletion of the stratospheric ozone layer, the enhanced greenhouse effect, and the loss of biodiversity. These global concerns framed our changing conception of environmental problems. With this went a new framing of the possible solutions, so that by the early twenty-first century the hope was that money would be made, not by polluting, but by cleaning up our environment and being more efficient with our resources. New tasks for environmental planning were proposed: restructuring the tax system towards eco-taxes and creating tradeable rights to aid environmental protection, as well as making environmental regulation more rigorous and stringent.

So we, in developed countries, have lived with an expectation of significant state action around the environment for well over 100 years. Environmental politics—meaning collective action focused on the state around the goal of environmental protection—has become an increasingly mainstream activity.

Environmental policy—meaning the package of government statements, strategies, and plans concerned with the environment—involves more and more governmental units, from the Treasury to the Foreign Office to the specifically named environment departments, from central to local government, from international agencies to national and regional quasi-governmental agencies. And our view of the state in relation to the environment is such that we now expect 'environmental planning'—meaning the whole process of state involvement with the environmental from formulation of strategies to their implementation.

The question is, how should we look at this environmental planning? Two main approaches have developed. The first looks at environmental planning as a rational justified state activity. There is a well-developed body of planning theory along these lines, together with substantial policy documentation that takes this approach. The second focuses instead on planning as a way of handling conflicts over environmental issues and building consensus on how to deal with these issues. Much academic literature on environmental planning, both critical and normative, falls into this category. This book suggests the value of an alternative approach, one based in analysis of discourse within environmental planning. Such an approach is critical and questioning in terms of the self-presentation of policy and decisions. It is premissed on the idea that the terms in which such policy and decisions are expressed are highly significant. In short, it argues that discourse matters. The full theoretical framework for this argument is presented in Chs. 2 and 3, but this chapter introduces the discourse approach by unpacking, deconstructing, or (more simply) questioning some of the key assumptions and concepts of contemporary environmental planning. In particular, it concentrates on the aspects identified in the title: rationality, conflict, and consensus within environmental planning.

Environmental Planning as Rational

The very term 'environmental planning' carries with it connotations of an active role for the planning agency (usually assumed to be part of the state), doing something to shape our use of and interaction with the environment. It is thus both broader and more specific than the term 'environmental policy'. Planning implies statement and action, more than just the framework for that action. It raises issues of how different actors engage with each other in order to achieve (or prevent) change. It involves issues of outcomes (such as decisions or detailed programmes) and even impacts (on the environment), rather than just policy outputs (which may take the form of legislation or a plan) (Gouldson and Murphy 1998). And yet planning is more specific in that it carries with it the idea that there will be action on the part of a state agency, responsibility borne by that agency. Environmental policy may result in a framework in which action and responsibility are dispersed among private organizations; this is not planning. Planning is something that state agencies do, although never in isolation, and it therefore involves the legitimacy of the state to act. The connection between legitimacy and action within the whole process of environmental regulation is one of the key foci of this book.

Central to the legitimacy of environmental planning is the idea that it is a rational process, pursued in the public interest using intelligence and knowledge, and grounded in a broad-based acceptance. This is an ideal, but an ideal that is widely accepted. To take a recent example, the British foot-and-mouth epidemic of 2000 can be seen in this light. This virulent virus spread rapidly among the British sheep and cattle stocks, threatening not only the health of specific herds but also the British export market in meat and meat products. In an attempt to maintain Britain's foot-and-mouth-free status, the government embarked on the traditional policy of culling infected and potentially infected animals. These animals then had to be disposed of by transport to rendering or incineration plants, burial, or open-air burning. Problems of delays in slaughter times and rising stocks of carcasses led to a vigorous public debate in the media and political circles. Should a policy of vaccination be adopted instead of or as well as the cull? What was the best method of disposal of the carcasses? Did open-air burning carry with it severe air pollution risks? Did disposal by burial affect the water table?

At the same time, movements in the countryside were limited, and many access areas, as well as all footpaths, were closed for a time. This raised further questions. Had the impact on tourism and the rural economy adequately been taken into account? Was protection of the export industry in meat and meat products worth the adverse impact on tourism? Were people being sensible in their response to the countryside restrictions given that only a minority actually ventured onto footpaths when visiting these areas? All these questions speak of the assumption that such a policy should be rational, justifiable, and able to command the support of those who have thought logically about the issue. In this line, the response of government ministers—when questioned about the problems arising with the disposal of carcasses—was to invoke a risk assessment, a quintessentially rational approach to the handling of different policy options.

However, study after study of environmental planning has shown that practice actually falls far short of this ideal. As Fischer (2000: 43) points out in his analysis of citizens, experts, and environmental planning: 'in the "real world" of public policy there is no such thing as a purely technical decision'. More than this, these studies have shown that the rationality of planning has been used to legitimate planning processes and outcomes even though there is a shortfall between practice and ideal. The question that is raised is how the burdens and benefits of environmental planning fall on different groups within society and how such legitimation may hide the distributive pattern. From this perspective, planning is about the conflict and congruence of interests and the associated play of power. Outcomes are the result of more and less powerful actors engaging with each other in specific contexts. The rationality of the policy process is shown to be an illusion, a cloak for the operation of power (Flyvberg 1998*a*).

Turning to the foot-and-mouth episode again, a very different story can be told. This sees the policy response irreducibly in terms of the different interests involved. Among these can be counted: the farming industry, the rural tourist industry, the Labour government, the Ministry of Agriculture, Fisheries, and Food (now Department of Environment, Food, and Rural Affairs); and the Environment Agency. Each of these actors has its own agenda to pursue. For economic actors, this relates to financial gain and loss. For some farmers, culling was preferable to vaccination as they received compensation for the former but not the latter. For the tourist industry, whether the preferred option was vaccination or a complete cull was immaterial, just as long as the matter was resolved quickly, the restrictions on countryside access and the damage to the image of the countryside were ended, and tourist revenue could start to flow again. But for other actors, the agenda was measured in other currency. For the government, the desire to remove any sense of an unresolved crisis before calling a general election was a key factor. And for the government departments and agencies there were issues of status, expertise, and autonomy in pursuing their particular approach. As the whole episode waxed and waned, the influence of these different actors changed. Detailed analysis of the interplay of interests would provide an account of environmental planning as the exercise of power and as the dominance of certain interests over others. It would highlight the inevitability of conflicts as well as the limited potential for co-operation between parties.

Analysis of environmental planning, therefore, needs to recognize that the legitimation of planning involves rationality claims. The key to understanding these claims is to see them as socially constructed, as constituted through discourse. The essence of claims to rationality is that such claims embody certain assumptions about what is the appropriate, even logical course of action. It suggests a way of thinking that will lead to the best outcome. These assumptions and ways of thinking are norms, the accepted values of what constitutes rationality. They are expressed through discourse, the use of language to express thought, intentions, values, and courses of action. It therefore follows that a close attention to the detail of discourse will enable us to understand the claims of rationality more fully, the ways in which they relate to institutional norms and the linkages to processes of policy legitimation. This will be examined further in terms of the theoretical discussion of discourse, drawing both on the social constructivist approach but also the detailed analysis of institutions and language use. It will also be examined through detailed analysis of specific types of rationality claims as they manifest themselves in environmental planning.

Conflict and Consensus in Environmental Planning

Critiques of current practice (such as Fischer 2000 and Flyvberg 1998a) have not just stopped short at unpacking the legitimacy claims of planning, though, but gone further in suggesting new modes of planning. These have tended to locate the main source of the problem in the lack of transparency and openness of state planning processes. Indeed the rationality claims are seen as a way of disguising this lack. By implication, the solution lies in opening up planning processes to greater participation by a variety of actors, often now termed 'stakeholders'. In the environmental domain, the circle of stakeholders is drawn very widely, following the view of the environment as a public or common good. But however this circle is defined, environmental planning becomes more than just a state activity. It becomes the privilege and responsibility of a wider group; rather than an aspect of state or government activity, it becomes a mode of governance.

Environmental governance is characterized by the wider involvement of interested parties, both in policy formulation but also implementation processes. Partnership mechanisms are often adopted to facilitate such implementation, the central idea being that the state does not have the capacity on its own to achieve many of its policy objectives (C. Stone 1989). Successful planning, therefore, requires the joint involvement of multiple actors. In this way, more inclusive government in the form of governance is legitimated on the basis of improved policy delivery (Rydin and Pennington 2000). But it is also legitimated in democratic terms, that is, the right of parties to be involved in decisions that will affect their lives. Governance is about participation and empowerment as much as delivery and effectiveness. Stakeholders are defined in terms of rights and the impact of planning, as much as their role as de facto planners.

Arguments for greater democracy, deliberation, and collaboration within planning processes all take this form. They emphasize the potential for building consensus as a basis for environmental planning decisions, a more inclusive consensus than has been seen so far. Such an emphasis involves the suggestion that this would be a better basis for planning practice, one that would less systematically disadvantage specific groups. Environmental planning would be less in need of the false legitimacy of rationality if such consensus could be achieved. But the shift from the analysis of environmental planning practice to environmental governance prescriptions does not diminish the importance of understanding legitimacy claims. On the contrary, it demands an analysis of these alternative claims, based on inclusivity, empowerment, and, in Habermas's terms, communicative rationality. These will be discussed further at several points in the book.

One interesting feature of these prescriptions is that they have an explicit substantive focus on discourse, not limited to the analysis and deconstruction of rationality claims. Rather discourse becomes something that is manipulated within the planning or governance process, for example in order to achieve a consensus. If discourse is seen as mediating all aspects of social life, including the interaction of interests, then using discourse creatively becomes a way of overcoming conflicts and creating agreement. However, discourse does not just mediate between interests, which are somehow defined outside the realm of discourse. Interests, the existence of conflicts, and the potential for consensus are also socially constructed which means that discourse is implicated in creating such conflict and consensus. To be clear, this does not mean that conflict and consensus only exist at the level of perceptions. Just as relationships between interests are not outside the realm of discourse, neither do they exist only within that realm. The key to understanding this, as is explained in Ch. 3, is the concept of institutions, social entities that both have an inherent discursive dimension and also present actors with incentive structures that are not endlessly malleable.

To illustrate the nature of the institutional discourse analysis that is presented in this book, two examples will be briefly presented. These relate, first, to the way in which common interests are discursively constructed and, second, to one specific case of the institutionalization of conflicts.

The Discourse of Common Environmental Interests: the Brundtland Report

> We are common shareholders in the progress of humankind.
>
> President Clinton in an address to the United Nations
> 27 September 1983 (quoted in De Luca 1999: 36)

Taylor and Buttel (1992) have argued, in relation to global issues, that 'We know we have . . . environmental problems, in part, because we act as if we are a unitary and not a differentiated "we".' The representation of the environmental agenda as somehow involving a common interest is now so widespread as to be practically ubiquitous. But the current association of a common interest with the environment is heavily tied up with the emergence of a global environmental agenda. For it is the global issues—ozone depletion, global warming—that are seen as the epitome of environmental impacts that affect us all. This line of argumentation has then spread to many other environmental issues, involving a smaller subset of 'us all' but nevertheless seeing that smaller subset as having a common interest. This is partly due to the particular way that the global agenda has been framed under the banner of sustainable development. Dryzek (1997: 123) calls the sustainable development discourse 'arguably the dominant global discourse of ecological concern' and Hajer (1995) has identified it a major success story. As Dryzek (ibid. 129) points out, one feature of this discourse is that, while it has its origins in global concerns, 'it does not rest at that global level' but has become an issue for regional and local levels also. It is thus applicable to and supportive of the notion of common interests in many arenas. And, of course, environmentalists have been keen to take up this line of argument since, like all lobbyists, they wish to persuade us that their values and interests are held in common with the larger part of the general populace, and are not a sectional interest or concern. But how does this line of argumentation work?

One of the key arguments of this book is that it is necessary to examine the exercise of discourse in some detail in order to understand how it works. And to achieve this, it is necessary to have a method of discourse analysis. The chosen method is rhetoric (following Myerson and Rydin 1996*a*, but see also Dryzek 1997 and Hood 1998). Rhetoric works by seeing discourses as implicit or explicit arguments and looking for the devices that make these arguments more or less effective. Among such devices are metaphors, synecdoche, metonymy, association, ethos, and mode of closure. The Appendix provides a fuller account of the method and these devices but, briefly, they can be summarized as follows:

- ethos, the personification of the speaker
- pathos, the mood music of the piece
- logos, the path of argumentation using various tropes such as:
  - metaphor, describing one thing in terms of another
  - synecdoche, taking the part for the whole
  - metonymy, one thing standing for another
  - irony, literally saying one thing and clearly meaning its opposite.

Rhetoric will be used at various points in the book but a closer look at the contemporary discourse of common environmental interests provides an illustration of its use.

A key text, perhaps the key text in contemporary arguments for common environmental interests is the Brundtland Report (WCED 1987). A summary of the discourse of this text is provided in terms of a 'rhetoric line' or line of argumentation in Fig. 1.1. A starting point for the rhetorical analysis is the metaphorical dimension of the discourse. While Dryzek (1997: 131–2) points generally to the use of organic metaphors in sustainable development discourse, in fact the metaphors of the Brundtland Report are more specific, drawing on parallels for the environment with a home for the family or a habitat for a species. These invite personal association, the identification of the reader with the broader issues being discussed. But further, such metaphors make disagreement and an emphasis on conflict more difficult. Everyone knows that families can be sites of conflict but their use as metaphors in an exhortatory, normative discourse, such as this, is meant to highlight the positive, nurturing side of family life, the family as safe haven. It also emphasizes the lowest common denominator of all human, or even just lived existence; we are all parts of families, all have homes, all need a habitat. This again discourages identifications that emphasize conflict or even difference. The types of

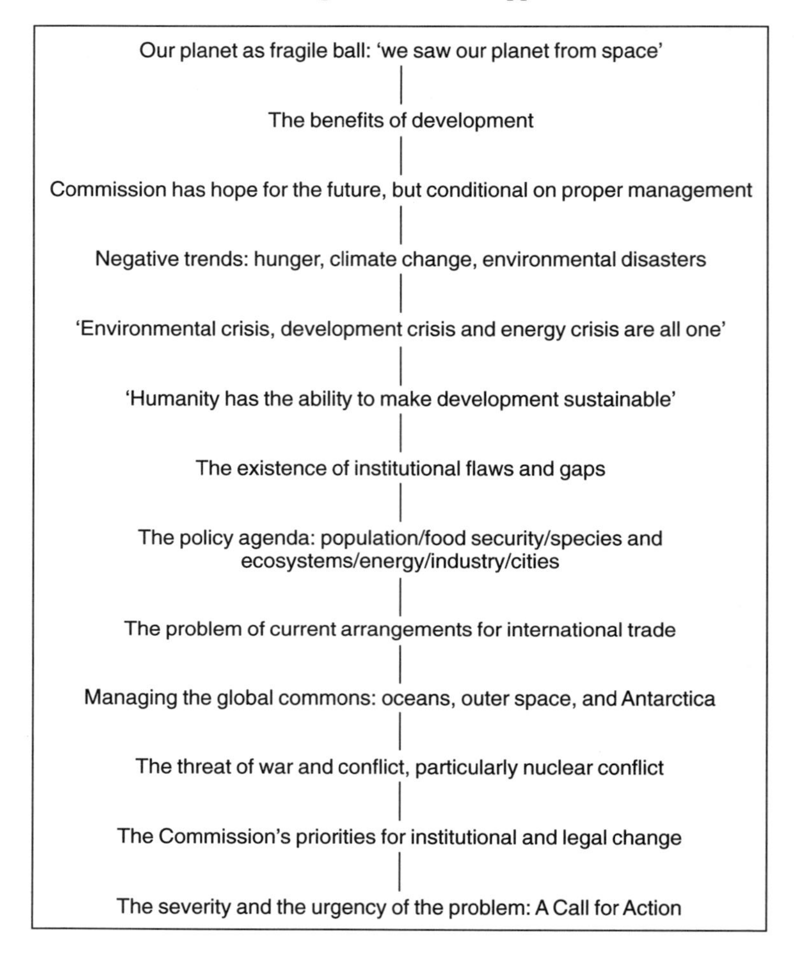

Fig. 1.1. A rhetoric line for *Our Common Future*

synecdochic reasoning used—where the part is used to stand for the whole—reinforce this further. Throughout these texts one finds examples of the argument making the leap from a focus on a specific group to a conclusion about humanity, the human species, people in general. These types of rhetoric play down the existence of competition and conflict (which Dryzek also notes) and play up the potential for consensus.

Again the ethos or personification of the speaker of the discourse reinforces this line of argumentation. The characterization of the speaker or rhetor is one of leadership, speaking to all people and for all people in an enlightened and visionary manner; the first-person plural is used repeatedly in all its forms. This often implies a superior vision of the challenge and of the way forward (see also

Myerson and Rydin 1996*a*: 120–4); a sustainable development focus enables one to see the present and the future in a 'better' way. But this is not the vision of a privileged few, for 'the onus lies with no one group of nations' but with 'the entire human family of nations' (WCED 1987: 22). This point is made easier to support by the use of metonymy in which 'the people' and 'the planet' are used interchangeably, so that the leader of people can effectively speak for the planet. This shifting between the social and the physical, also noted in the metaphors used, adds to the authority of the discourse since no mediation is apparently needed (for example, by scientific authority) to move between the realms. It also occludes any potential tension between socio-economic dimensions (the development side) and the ecological (the sustainability side); this is important rhetorically as the absence of any *necessary* tension is central to the argument of the discourse. Sustainable development is the manifestation of a genuine win-win scenario, which is blocked by outside opposition rather than internal tensions. As Dryzek emphasizes, it necessarily implies progress.

In these ways it becomes difficult to conceive of disagreement between people on the issue of sustainable development and, by implication, with the concept of sustainable development itself. As Myerson and Rydin (1996*b*: 23) noted, it is not plausible to use the *presence* of sustainable development as an argument *against* something in the way that conservation, say, can be used both negatively and positively. Rather, if sustainable development can be demonstrated, it must be a positive feature. The type of closure to argumentation that is used adds strength to all these argumentative devices by emphasizing urgency, crisis, and the absence of any alternative path. The leadership provided by the discourse is both essential and the right way. This, again, precludes any argument against sustainable development. In these ways, sustainable development becomes a very powerful argument for a common interest in global environmental issues, an argument that is based on the lack of inherent disagreement between peoples and that cannot itself be readily disagreed with. However, this does not mean that a discourse of common interest is always persuasive. While in some circumstances discourses reinforce each other through rhetorical complementarities, often they compete with each other for dominance in providing the preferred construction of a situation or event.

Furthermore the prevailing discourses need to be appropriate to the institutional context. This means that a discourse of common interests will be more relevant where that institutional context supports actors' decision-making on the basis of common environmental interests. For many, the notion that aspects of the environment are irreducibly common is strongly associated with the definition of the environment as a public good. The defining characteristic of a public good is usually taken as jointness of supply. Public goods are jointly supplied with regard to a given public because: 'consumption of any unit of the good by any member of the public in question does not prevent any other member consuming the same unit' (Taylor and Ward 1982: 351). Other defining features are the indivisibility of the good and its non-excludability. In addition, Dowding and Dunleavy (1996) identify two other associated features: discreteness and crowdability. These features are often spoken or written of as if they were inherent features of the good, a necessary result of its physical nature. And yet, as Dowding and Dunleavy (ibid.) emphasize, the extent to which these five criteria apply depends on the prevailing institutional context and, in particular, the way in which technology applies within it.

Take, for example, the central concept of joint supply. Whether the consumption of a good by one diminishes the amount available to others depends on the conditions under which that good is provided and, in particular, it depends on how that consumption is managed. Management here need not mean rationing; there are the subtle ways in which countryside and recreation managers direct visitors to different areas within a country park, say, in order to accommodate a larger number of visitors with minimum environmental degradation. Similarly technology can ensure that a particular resource, such as water, is managed so that the threshold within which joint supply applies is increased. Dowding and Dunleavy similarly explore the institutional bases of the other defining characteristics of public goods. Indivisibility depends on how the good is defined; non-excludability depends on how access is defined and managed; and so on.

Therefore, in all these respects, the definition of an environmental good as a public good is a function of institutional arrangements, of the current organization of an environmental good or service, and not timeless dimensions of the environment. This means that there are choices involved, choices about the type of institutional arrangements that should be adopted for any specific aspect of the environment and, therefore, choices about who benefits from these arrangements. These choices go alongside the use of language in constructing a view of the environment as held in common or not. Institutional arrangements involve more than just organizational arrangements, more than just discourse; they imply both. An institutional discourse approach recognizes both these aspects.

Constructing Environmental Conflicts: Media Institutions and the Environment

Just as common environmental interests are discursively constructed and institutionally embedded, so are conflicts of interests concerning the environment. It is not just the case that conflicts are, in some sense, material and 'real'; while common interests are a form of manipulation. Both depend on construction processes and institutional arrangements. To illustrate this in relation to environmental conflicts, one specific set of institutional arrangements—the media—will be discussed.

The role of the media in the presentation of environmental issues is a significant one, that has been the subject of much research in recent years. The significance of media coverage of the environment is, of course, extensive and only one aspect will be considered here. This concerns the considerable evidence that the ways in which environmental stories are constructed within the news media can foster a climate of conflict rather than co-operation. There are three acknowledged institutional influences on the construction of news stories: the wishes of owners of the media; the perceived demand of the market for news stories, the audience; and the bureaucratic pressures on and normal working practices of the creators of news stories (Hannigan 1995; De Luca 1999). Most accounts of the mass media emphasize the role of organizational routines in generating news through timesaving and repeated practices (Hannigan 1995: 59–60, Sachtsman 1996: 251, Schlechtweg 1996: 257). These include: routine runs of predictable stories, repeated sources and loci of news stories, the use of existing formats for telling stories and for new 'angles', and conventional frameworks that filter out unacceptable stories and orient acceptable stories. This conclusion is reinforced in work that looks specifically at the environmental domain (Burgess, Harrison, and Maiteny 1991; Hansen 1991, 1993; Lacey and Longman 1993).

The nature of these routine stories in relation to coverage of environmental stories is amplified by Chapman *et al.* (1997: 40–1). Typically, they emphasize the importance of conflict and tension, sometimes with the tag of novelty. Individuals are a common focus, usually characterized as a 'good guy' or 'bad guy'. There is also a preference for stories that can be visually treated. The theme of how conflict underpins news stories runs through most of the literature. In their study of journalistic norms Miller and Parnell Reichert (2000) found that conflict is a driving force in the selection of stories. S. Douglas (1992: 268) also emphasizes the role that conflict plays within the presentation of mass media news: 'media need societal conflict to function; they profit from the conflict'. In his study of the frameworks of interpretation and evaluation in the case of a television programme on logging on the northern coast of California, Schlechtweg (1996: 274–5) refers to 'the binary, oppositional logic that informs the frame' and concludes that this 'constructs a cleavage' (in this case between radical environmentalists and 'regular people') and 'reinforce[s] the original defining opposition'.

Analysis of the language of environmental coverage has noted that, not only are the stories set up in terms of opposition between parties, but they extensively use conflict metaphors. Rydin and Pennington (2001) found this in their research on local press coverage of environmental issues and De Luca (1999) makes similar comments in his analysis of the media and environmental risk. He also points out that discourses around sport, sexuality, and politics are regularly imbued with conflict metaphors, making rhetorical links out from environmental discourses. Meanwhile Viehöver (2001) notes that media coverage of waste management in Germany increased in the period up to mid-1999 because the narratives adopted by the dominant discourse coalitions coincided with the media's preferences for dramatic plots, bad news, and conflicts.

In the environmental arena, it appears that the institutionalized preference for conflict stories may actually skew coverage of environmental activism that is seeking to be non-confrontational and take the debate beyond the frame of conflict interests. De Luca (ibid. 90) argues that it is the current fate of the environmental justice movement to be covered in the press as simply a form of NIMBY (not in my back yard) anti-development movement. Attempts to broaden out debates are also constrained by the media's preference for coverage of events, rather than sustained critiques. Furthermore, prevailing patterns of coverage are likely to encourage certain forms of protest action, hence the tendency for groups such as Greenpeace to stage media 'events' in order to provide a platform for their broader message; whether the broader message ever gets sufficient airing in this way is a moot point. Miller and Parnell Reichert (2000) also argue that, while certain forms of protest were presented as illegitimate in the media, nevertheless a degree of protest was needed actually to obtain coverage. For example, take the media coverage of the anti-roads and anti-airports protesters, who have used their own bodies in direct action against development, living up in trees or in tunnels beneath development sites (Wall 1999; Wykes 2000). This is a clear-cut story of conflict between the developers and the protestors, sometimes resulting in physical conflict. The protests themselves are highly photogenic and, in terms of the discussion above, lend themselves to media coverage in the press and television. The protestors are engaging in collective protest action in the anticipation of such media coverage; the coverage is an additional incentive—over and above the other factors shaping the protest—to collective action.

In their study of local environmental policy and press coverage in Devon and Sussex, England, Rydin and Pennington (2001) found that there was a relationship between local newspaper coverage and collective action. Here it was argued that the expectation of positive and, more importantly, significant press coverage could act as an incentive to collective action. This was particularly the case with NIMBY coverage although, to a lesser extent, it was also a dynamic at work in coverage of positive collective action for local nature conservation. However, in such cases, stories were less likely to be given prominence, as collective action for nature conservation is essentially a 'feelgood' story and not one based on conflict. This brief excursion into the rich field of environmental coverage by the media highlights again the importance of social institutions in shaping the construction of environmental processes.

Outline of the Book

To recap, this book explores the issues of the construction of rationality within environmental planning, the role this plays in legitimating planning practice, the connection with the interplay of interests constructed as situations of conflict or commonality, and the scope for building alternative procedures and processes, which empower disadvantaged groups and potentially build policy consensus. In doing so, the analysis seeks to set the construction of discourses alongside the interplay of interests and to explore the specific role that discursive processes fulfil. It aims to identify the scope for agency, as well as recognizing the context for and constraints on that agency. It sets a concern with discourse within a broader institutional perspective. This enables the role of language and discursive processes to be analysed in terms of broader social and cultural processes, as well as the interaction of specific interests. Finally, it provides a means for seeing how discursive acts are involved in the very fabric of everyday environmental planning.

Chapter 2 considers alternative theoretical approaches to analysing discourse within environmental planning. This encompasses the social constructivist paradigm, specific contributions from political science, and applications of the Foucauldian approach, with its view of discourse as power, as well as the Habermasian perspective. The discussion suggests the need for some theory-building to support critical analysis of discourse and agency within the environmental policy process.

Building on these arguments, Ch. 3 then sets out an alternative approach, beginning from an institutionalist standpoint and building in particular upon Elinor Ostrom's work. The institutionalist standpoint is seen as providing an ideal bridging point for considering discourse in terms of structure and agency, rationality claims, and the interplay of interests. It enables consideration of the incentive structure facing actors to be combined with active decision-making on the part of those actors. Through its emphasis on norms, values, and working routines, it suggests a central role for discourse in establishing and maintaining such norms, values, and routines. To date, however, this discourse role has not been elaborated. The chapter, therefore, outlines a three-level framework in which discourse is approached through structuring contexts (the term 'discourse' is thereafter reserved for this level), arenas for communication, and the active use of language through discursive strategies. This provides the framework for subsequent analysis of actors and discourses within different dimensions of environmental planning.

In Ch. 4 the analysis considers how conflicts within environmental planning can be handled, ranging from reframing strategies, through negotiation and bargaining to consensus-building approaches such as collaboration and deliberation. Considerable use is made of reported research to draw out the specific role that discourse can play. The emphasis is on identifying the potential of these different techniques and approaches but also their limitations. The conclusion identifies how institutional change has to go alongside discursive strategies in achieving resolution of conflicts and developing a more common approach.

Chapter 5 returns to the key theme of the self-presentation of environmental planning as a rational process. It analyses the way in which this self-presentation is supported by a specific procedural discourse and then links this with the interests of actors within environmental planning bureaucracies. Setting the interests of bureaucrats, the notion of rational policy work, and the analysis of policy discourses alongside each other provides a more complete understanding of how environmental planning works. The institutional context, actors' interests, and the discursive dynamics are then examined further in terms of the work of plan-making, regulation, and public participation. This links back to the discussion about more inclusionary forms of decision-making in Ch. 4.

Chapter 6 continues the theme of discourses of environmental planning by considering the substantive rationalities that are used to rationalize or legitimize planning policy and practice. Three such rationalities are examined: scientific, economic, and communicative. Using specific examples, analysed through the medium of rhetoric, the potential and contradictions of these rationalities are examined. The linkages between these are also considered and linkages drawn back to the theme of inclusionary decision-making and planning expertise, covered in the previous two chapters.

Three case studies are then presented in Chs. 7, 8, and 9, which draw together the different threads of the analysis so far, emphasizing the ways in which procedural and substantive rationalities interact, and how the discursive dimension connects with the interplay of interests in each case. The case studies presented concern air pollution control and air quality management, planning for housing land, and Local Agenda 21. These cases have been chosen to emphasize again the themes of rationality claims, conflicts of interest, and inclusionary planning and expertise.

Finally, in Ch. 10 the prospects for developing a rationality oriented towards supporting and legitimizing a sustainable development policy is approached. This is discussed in terms of the rhetorical quality of such a discourse of sustainable development rationality and also the institutional requirements for embedding such a discourse in policy and planning practice. Through the successive analysis of these chapters, it is hoped that the reader will have a clearer understanding of the many-faceted ways in which discourse plays a role in environmental planning.

2

Discourse and Environmental Planning

> And the Lord said, Behold, the people is one, and they have all one language; and this they begin to do: and now nothing will be restrained from them, which they have imagined to do. Go to, let us go down, and there confound their language, that they may not understand one another's speech.
>
> (Genesis 11: 1–9)

The central focus of this book is the role of discourse within the processes of environmental planning, particularly as it relates to rationality claims and to conflicts and consensus within these processes. The book examines in detail the processes of agenda creation, legitimation, conflict mediation, and consensus building, as well as the more routine aspects of bureaucratic planning practice. It does so in the context of a broader concern with discourses, communication, and argument within planning processes. In this chapter the existing literature relevant to such concerns is examined, highlighting issues for further consideration and identifying gaps and weaknesses in current formulations. The starting point is the social constructivist approach, which 'reveals the hidden' and unpacks the 'taken-for-granted' within environmental planning, before moving on to the contribution of political science approaches and the major contemporary debate between Foucauldians and Habermasians.

Social Constructivism: Revealing the Hidden

The rise of social constructivism over the last two decades has been a major reason for the interest in the nature of environmental discourses and the importance of such discourse in shaping our understanding of the environment (Yearley 1991; Harrison and Burgess 1994; Hannigan 1995; Macnaghten and Urry 1998). To some extent, the growing acceptance of a social constructivist stance is just the result of academic disciplines tracking a new social phenomenon. There has been, as Lash, Szersznyski, and Wynne (1996: 1) put it, an 'unprecedented outbreak of environmental discourse'. But there has also been a broader 'discursive turn' within the social sciences, in which discourses—including environmental discourses—have been given a new significance.

At first, in the late 1970s and early 1980s, this was limited to a recognition that all social practice is mediated through language and, therefore, language as expressed through discourse must play a role in shaping those practices. By the late 1980s and into the 1990s this became associated with emerging analysis of late twentieth-century society as entering a postmodern period, characterized by the dominance of mass modes of communication, a concern with representation, and the rise of the self-reflexive subject (Connor 1989; Seidman 1994). While there remains much academic debate about the meaning of the terms 'postmodern/ism/ity', it is symptomatic of a sea change in society that a self-aware and explicitly expressed concern with discourse has become a feature of the behaviour of political parties, non-governmental groups, and the media.

As a result of these shifts, more disciplines have been ready to consider environmental discourse as part of their remit. This has also contributed to the rise of the social constructivist perspective. One key example is environmental sociology and Hannigan (1995) succinctly outlines the growth of this branch of sociology with its social constructivist agenda. Similar approaches can be found within cultural geography (Harrison and Burgess 1994) and environmental anthropology (Milton 1993, 1996). Meanwhile in the United States, practitioners of environmental rhetoric have found a new outlet for academic study in the growth of environmental texts (e.g. Killingsworth and Palmer 1992; Cantrill and Oravec 1996; Herndl and Brown 1996).

The basic premiss of the social constructivist approach to the environment is that social and cultural processes shape our perceptions of the environment and the ways in which we represent it through image and word. So the primary emphasis of social constructivism is on opening up the question of how we interpret and represent the environment, indicating a refusal to take the environment as a given that can only be depicted in a set way. A river is not just a river; it can be a subject for poetry, a habitat for water birds, a channel of transport, a barrier, a boundary, a hazard. It is, therefore, a relevant matter for empirical enquiry to investigate the different ways in which the environment or elements of it are constructed and how different groups in society construct the environment differently.

A further emphasis of this work—particularly by sociologists and anthropologists—is on the social nature of these construction processes. For while we each engage in a variety of individual acts that depict the environment, we do so in a social context. We draw on a limited set of communicative resources for our oral acts, those resources that are available within the society we occupy. The cultural context in which we find ourselves enables and constrains the possibilities for constructing the environment. In some societies, the 'environment' itself is not a significant construct (Chapman *et al.* 1997). And aspects that appear common to all societies—such as the existence of animals, for example—are in practice very differently constructed, as anthropological studies reveal. Compare the highly anthropocentric construction of animals in the British media with the very distinctive construction by the Mende tribe in Africa (Richards 1993).

Furthermore, in so far as we expect these acts to be understood and responded to, we need to ensure that the communication is not just sent but also received. This means that the generation of discourse must take into account the interrelationships between actors within specific contexts and society in general. Communication is a social act and few depictions of the environment are meant just for the benefit of the originator. So social constructivism gives equal weight to the 'social' and to 'construction' in its unpacking of the processes of generating and using discourses.

One particular strength of social constructivism is the way in which it questions the 'taken for granted' in contemporary environmental debates and encourages everyone to engage in such questioning. The relatively non-technical nature of many analyses within social constructivism, and the emphasis on the shock of recognition that comes from seeing environmental issues in a new way is an advantage, as it effectively democratizes the insights of the approach. Teachers approaching social constructivism may find students unresponsive to the approach when mapped out in theoretical terms but usually discussion of texts, new and familiar, will result in vigorous discussion and a ready identification of social construction processes, particularly with more media-savvy younger generations. There is a real and rewarding pedagogic role involved in exposing 'some very basic settled assumptions to critical examination', as Redclift and Benton (1994: 2) describe it.

But beyond this, the social constructivist approach claims that it is able to provide new knowledge by focusing on such discourses, mapping their variety, and understanding how they are formed in detail. There is the understanding of how an issue becomes an environmental issue and how it becomes a particular type of environmental issue, say a pollution issue affecting local ecosystems rather than a justice issue affecting local populations or a resource efficiency issue affecting businesses. This is the point that Hannigan (1995: 2) makes at the outset of his survey of social constructivism (or constructionism, as he terms it): 'Environmental problems do not materialise by themselves; rather, they must be "constructed" by individuals and organizations who define pollution or some other objective condition as worrisome and seek to do something about it.' As Dryzek (1997: 7) emphasizes in his study of the varieties of environmental discourse: 'environmental issues do not present themselves to us in well-defined boxes' with labels. Therefore, we need to understand how the environmental arena is cut up into the labelled categories that we daily face and use.

But this is not just a typological task. As Dryzek (ibid.) argues: 'This inquiry rests on the contention that language matters, that the way we construct, interpret, and analyze environmental problems has all kinds of consequences.' Hannigan (1995: 3) specifically points to the importance of social constructions for legitimating particular points of view and, therefore, sees social constructivism as helping to understand 'how environmental claims are created, legitimated and contested' (1995: 3). Adopting a social constructivist approach enables the analyst to examine the consequences of particular modes of social construction and of the ways in which environmental positions are thereby legitimated. It involves challenging the taken-for-granted and, thereby, implies a challenge to dominant modes of legitimation. This in turn opens the way for considering alternative positions and arguments. Given that particular groups will be associated with particular positions, the analysis can highlight the privileging of certain groups within environmental discourses and the marginalizing of others. These are clearly points of central importance of any study of the policy process and will be returned to at numerous points in subsequent chapters.

The legacy of modernism means that, in many societies, such legitimation is closely tied up with rationalization based on knowledge claims. The claims to authority that constitute legitimation go along with claims to rational, informed decision-making. This will be a central issue discussed at length in Chs. 5 and 6. For the moment, it is worth noting that rationality can be based in a range of claims. Within environmental planning, such claims usually take the form of knowledge claims; legitimation on the basis of the irrational or the dismissal of knowledge is comparatively rare. Social constructivist perspectives can highlight the knowledge claims of certain discursive positions. Particular attention has been paid to the ways in which environmental science justifies its knowledge claims and how this relates to the role of conventional science in legitimating certain policy approaches (Irwin 1995). Attention to this has also opened up the possibility of widening the range of knowledge claims within environmental debates. There has been a particular emphasis on lay knowledges (Irwin 1995; Wynne 1996) and on the knowledges of indigenous peoples as revealed through anthropological study (Milton 1993, 1996).

However, to perform these tasks social constructivism needs to be placed within an analytic framework. It does not, of itself, enable statements to be made about the exercise of power, legitimation, or rationalization. Social constructivism is an approach and a general methodology but not an analytic framework that will address such issues. To form such a framework requires some additional conceptual links. In effect, it requires a conceptualization of power, interests, and incentives. The discussion now turns to such approaches, beginning with those within the political science tradition who have considered issues of social construction and discourse, though not necessarily under these terms.

What about the Politics?

The concern within political science in 'revealing the hidden' is linked to the critique of pluralism as the dominant paradigm for analysing political processes.

Within pluralism, political action is overt, observable, and a response to explicit values, interests, and concerns. It leaves little or no room for understanding that the spaces for political action are shaped in such a way that certain paths are open and others closed. As early as the 1960s, Schattschneider (1960: 71) recognized the importance of these aspects and coined the suggestive term 'mobilization of bias' to describe how political institutions organize and allocate political space in order to manage conflicts. 'The crucial problem in politics, is the management of conflict,' he states. One way to achieve this is to organize certain issues out of political discussion, so that they never arise as explicit issues: 'Some issues are organized into politics, while others are organized out' (ibid.). In contemporary terms, this is about the nature of political discourse and how the political agenda is constructed.

Bachrach and Baratz (1962) develop this insight further, identifying two 'faces of power'. The overt face is seen in pluralist analyses; the covert face is seen at work in non-decision-making processes, whereby the political agenda is manipulated away from controversial issues and towards 'safe' ones. Here actors seek to create or reinforce 'social and political values and institutional practices that limit the scope of the political process' (ibid. 948). This involves investigating the dominant values, myths, established political procedures, and rules of the game alongside an assessment of which actors would lose or gain from this mobilization of bias. In a useful empirical study, Crenson (1971) applies this framework to an analysis of urban air pollution in two American cities, Gary and East Chicago, with very different traditions of pollution control, the former described by Crenson as 'lethargic', the latter as 'speedy'. He provides an account of how the political agenda is built in these two contexts, particularly through the rationality of political ideology. He identifies that, where economic development dominates the agenda, air pollution was organized out of the discussion. This is a clear case of powerful interests reinforcing their position through the political agenda.

A more generalized model of how political agendas are built is provided by Kingdon (1984), applied in many subsequent studies (see e.g. Wilder 1993 and Alm 1994 in the environmental domain). Kingdon's model involves identifying the key actors relevant to the context and identifying the resources available to them in order to influence the policy agenda or the range of acceptable policy alternatives, including electoral influence, economic power, and expertise, among others. This done, Kingdon specifies three streams running through organizations: the process of problem recognition; the formation and refining of policy proposals; and politics. These streams are independent but interact, leading to policy change that can be discontinuous or not. Thus, in Kingdon's account, the setting of the agenda and the devising of policy alternatives are not seen as distinct stages in a sequential policy process, but rather as different processes which interact.

The policy entrepreneur fulfils a key role within this model. This actor helps bring forward particular policy alternatives from out of the policy 'primeval soup'. Such entrepreneurs are characterized by personal traits of persistence, owning a claim to a hearing in policy forums, and a reputation for having good networks and/or negotiation skills. They are swayed by a set of incentives including the desire to promote personal interests, the desire to promote values, and the solidary benefits of being part of the policy process. However, Kingdon always qualifies the impact of any particular actor or set of actors by emphasizing the significance of ideas and quoting Keynes (1936: 383) approvingly. 'I am sure that the power of vested interests is vastly exaggerated compared with the gradual encroachment of ideas'.

Rather than identify a particular individual, Sabatier focuses on the role of coalitions in building policy agendas. Sabatier's advocacy coalition approach has much in common with Kingdon's agenda-setting analyses. They both stem from mainstream North American political science, with its roots in pluralism; look to the actions of visible political actors to explain the shaping of the political scene; place an emphasis on the way in which ideas are pursued within the political arena; ignore policy implementation; and have spawned much empirical research on the basis of their frameworks. However, Sabatier's framework is much more explicit in specifying the ways in which actors interact to achieve policy change. It is also of particular interest for the subject of this book since it has been used extensively to study environmental policy (Sabatier and Jenkins-Smith 1993, 1999; also J. Richardson 1994; Keck and Sikkink 1999).

The key concept of the 'advocacy coalition' is defined as comprising people from across governmental, non-governmental, and corporate organizations, who share a set of normative and causal beliefs and engage in a non-trivial degree of co-ordinated activity over time. A third party, a policy broker, sometimes mediates the strategies of these coalitions. Taken together these coalitions and policy brokers can explain policy change within a policy subsystem or domain over time-periods of a decade or more. It is possible to map the influence of advocacy coalitions because public policies and programmes can also be conceptualized in terms of values and perceptions in the same way as belief systems, and thus the trajectory of beliefs over time can be traced.

It is clear that the central factor bearing explanatory weight within Sabatier's work is the belief system. Each belief system can be specified at three levels:

- the deep core refers to basic ontological and normative beliefs;
- the policy core represents the coalition's basic normative commitments and causal perceptions across an entire policy domain; and
- the secondary aspects refer to narrower beliefs that do not extend across the entire domain, but relate to instrumental decisions and information searches necessary to implement the policy core.

While the deep core is very difficult to alter, the policy core can be changed by anomalous experiences of actors and the secondary aspects are relatively amenable to change. Hence much policy effort is focused on these secondary aspects. In particular, it is the policy core that acts as the glue holding the coalition together. In this way beliefs, expressed discursively, are a major influence on the ability of actors to work together. Deep core and policy core beliefs also act as important filters on new information or research. This is of significance as one of Sabatier and Jenkins-Smith's main purposes is to understand the role that technological and scientific expertise play in policy learning and, hence, change.

The key claim that Sabatier and Jenkins-Smith make is that empirical research has yielded a considerable body of evidence for the existence of advocacy coalitions and their stability over time. There has, however, been some debate about the basis for coalition building. Sabatier and Jenkins-Smith (1999: 138) admit that they had assumed that actors holding similar beliefs would act in concert. In contrast, rational-choice critics have pointed to the transactions costs involved in maintaining coalitions, the difficulties of finding policies that avoid distributional conflicts within the coalition, and the dangers of free-riding (Olson 1969; see also Ch. 5). Rather than take these points fully on board, Sabatier and Jenkins-Smith prefer to rely on a model of the individual actor that emphasizes beliefs as perceptual filters. Therefore, they argue that the common beliefs of those within advocacy coalitions will be a causal factor in reducing the costs of co-ordinated action and help create and maintain coalitions (ibid. 139).

One common feature of these political-science approaches has been the emphasis on actors and their role in shaping political agendas. Political action occurs in the context of ideas, beliefs, or ideologies, but there is little detailed attention given in any of these accounts to the role that language can actually play in shaping agendas. Kingdon talks of problem definition being affected by the use of symbols, values, the comparisons that people make, and the categories they use. This draws on a social-constructivist paradigm and is suggestive of a discourse analysis, but it is not pushed any further. Kingdon tends to associate the realm of ideas with 'superior argumentation' and doesn't discuss how the substance of ideas is related to their discursive presentation.

Neither does he allow for the way in which ideas affect the presentation and perception of interests or for the way in which interests affect the presentation and perception of ideas. This becomes more apparent when the criteria for 'survival' in the primeval soup of policy alternatives are set out: technical feasibility; value acceptability within the policy community; tolerable cost; anticipated public quiescence; and reasonable chance for receptivity among elected decision-makers. These are not neutral objective criteria; they are all socially constructed concepts. Their importance as filters in the policy process suggests that this is another aspect of the way in which the overall process is discursively created. However, Kingdon's approach does not extend his analysis beyond the explicit concern with the policy agenda into the more general issue of the social construction of policy discourses within institutions.

Similarly, in the advocacy-coalition approach, little is said about the actual role that communication plays in the policy process, even though the communication of beliefs is central to the processes at issue. Language is implicitly theorized as transparent and beliefs are unproblematically conveyed. And yet, at various stages, the importance of the expression of beliefs is hinted at. For example, in their more recent formulation of the approach, Sabatier and Jenkins-Smith (1999: 134) argue that purposive groups are 'more constrained in their expression' than material groups. As mentioned above, in debating the advocacy-coalition approach *vis-à-vis* rational choice, they argue that perceptions of benefits are a relevant factor in facilitating co-ordination. Again, they refer to strategies of portraying opponents as 'devils' and to the importance of the 'communication' of a plan (ibid. 140). Yet, in all these cases, there is no consideration of how the discursive resources available for expression, communication, and even perception itself may have an influence (de Leon 1999: 27).

One analyst who has taken this discursive dimension of politics seriously is Edelman (1988). Edelman's work addresses the policy agenda in general terms and sees politics as a whole in terms of its construction as a spectacle. Drawing on the terminology of Barthes, he seeks to demonstrate how terms such as 'social problem', 'political enemy', or 'leader' have a reality as signifiers with a range of meanings for different audiences. He also shows how the identity—including the self-identity—of political actors is constructed (ibid. 2): 'their [political actors'] actions and their language create their subjectivity, their sense of who they are'. Furthermore 'people involved in politics are symbols to their observers' and he discusses the role of identity construction in conflict management and the operation of bureaucracies. Much of this is highly innovative and insightful and will be drawn on later in the book. However, it remains largely untheorized, although a tentative connection is made to the key institutionalist writers March and Olsen and their garbage-can theory of politics (ibid. 21–2 n. 5).

Perhaps as a result, the conclusions to which Edelman is drawn are rather functionalist in nature. For Edelman the discursive construction of politics is all about the way in which entrenched interests and structures are maintained. He is interested in the way that the language of politics maintains the status quo. For example, he identifies how political constructions can mute conflicts of interest, reassure victims, and help moderate the intensity of social conflict (ibid. 14), and how ambiguities in political debate can foster ambivalence and acquiescence. Thus he argues (ibid. 19–20): 'The construction of problems and of reasons for them accordingly reinforces conventional social cleavage.' Furthermore, he shows how political language maintains the authority of those already in charge: 'The language that constructs a problem and provides an origin for it is also a rationale for vesting authority in people who claim some kind of competence' (ibid.). This allows little space for strategic action on the part of actors or for such action to lead to change. He tends to deny the possibility of interest-oriented strategies and conceives of rationality on the part of actors as 'ideology in the name of reason' (ibid. 109).

But, more significantly, this constrains the potential for social and political change to occur, and limits it to the non-discursive realm. As he states (ibid. 130): 'The dissemination of new political terms, concepts, and phrases without concomitant change in material conditions can only reinforce the old tensions and premises.' Change will, therefore, have to be based in material conditions and not in the discursive realm. This means that Edelman is essentially reducing political language to the pursuit of material interests; while he argues that language is an important way of evoking meaning for actors, that evocation 'takes place only as a function of a specified material and social condition' (ibid. 9). This is an account that does not allow a distinctive role for language within the policy process, other than as a reflection of the play of power and interests. As will become clear, this makes a linkage between Edelman's approach and the broader framework of Michel Foucault and his followers.

But reducing language to power is not the only alternative. One political scientist who has sought to incorporate a concern with language into an actor-centred account of politics is Riker. Riker (1986) has coined the term 'heresthetics' to encapsulate the art of political manipulation and sees this as adding a fourth dimension to the traditional trilogy of classical arts: logic, grammar, and rhetoric. He distinguishes rhetoric from heresthetics by separating out persuasion (the realm of rhetoric) from manipulation, where strategic value takes precedence over persuasive value. The aim of the heresthetician is to structure the decision-making situation to the speaker's advantage and the respondent's disadvantage (ibid. 8). This formulation places actors' interests centre-stage and links a discursive strategy to these interests. However, the actual term of heresthetics is constructed to extend across *and* beyond discursive strategies; it is not a distinctive way of using language, neither is it distinct from language-based strategies. For, while Riker claims that heresthetics is based on the use of language to shape facts, give reasons, and describe situations, he goes on to identify three main forms of heresthetics, not all of which are linguistic.

These three forms are agenda control, strategic voting, and manipulation of dimensions. Clearly strategic voting is not primarily a discursive strategy (although like all action it will have a discursive dimension: how the voting intention is conveyed, etc.). Agenda control and manipulation of dimensions are discursive; of these, agenda control has been dealt with, in more detail, by Kingdon, and Riker himself puts the emphasis on manipulation of dimensions. By this term, Riker means the shaping of a particular political issue, what aspects are considered and their relative priority. This can underpin political alliances by defining a proposal around which actors can group. It can alter the dynamics of political debate by redefining the key points for consideration. However, one interesting emphasis that Riker brings to the analysis is the self-perpetuating nature of the manipulation of dimensions: 'once performed it does its work without further exertion by the heresthetician' (ibid. 151). This is interesting because it suggests the possibility of an account that allows for actors undertaking discursive strategies and also discursive structures themselves having a distinct influence. In this way, Riker can be placed alongside the other great corpus concerning language and politics, the work of Jürgen Habermas and his followers. The discussion of Foucauldian and Habermasian approaches constitutes the remainder of this chapter.

The Discursive Critique of the Foucauldians

> [Foucault's] life's work has been an attempt to catch what the present was telling him over the din of the past still echoing in his ears.
>
> (Sheridan 1980: 196)

Foucault can be considered the ultimate social constructivist, in the sense that his perspective (in all its variants) derives so directly from the importance of social constructions organized into discourses. In particular, Foucault and Foucauldians are associated with the linking of the concepts of discourse and power. For Foucault (1984: 110), power is exercised through the presence and use of discourse: 'discourse is the power to be seized'. Such power is not a resource vested in actors nor even the generalized effect of a system of domination exercised by one group over another. This is a relational conceptualization of power. Power is pervasive and affects all within society; it exists within every social relationship. Two quotes from Foucault illustrate this:

> Power is everywhere: not because it embraces everything, but because it comes from everywhere . . . power is not an institution, nor a structure, nor a possession. (Foucault 1978: 93)

> When I think of the mechanics of power, I think of its capillary form of existence, of the extent to which power seeps into the very grain of individuals, reaches right into their bodies, permeates their gestures, their posture, what they say, how they learn to live and work with other people. (1977 interview with Foucault quoted in Sheridan 1980: 217)

The analysis of power centrally involves the analysis of discourses. Discourse and power are both defined in terms of each other.

Foucault pursued this approach through a number of projects and in a variety of ways. In particular, he sought to examine the 'three great systems of exclusion': forbidden speech, particularly in relation to sexuality; the imposed division of madness, including the use of specialized built institutions; and the will to truth, meaning the categorization of knowledge and the role of knowledge institutions. Knowledge, in particular, was seen as an integral element of power relations, so much so that the two terms are often bracketed together: 'knowledge/power'. Analysis took the form of identifying the principles of sanctioning, exclusion, and appropriation of discourse, which Foucault termed critical analysis. Centrally, historical analysis was used to specify how these discourses were generated, formed, and sustained; this is termed 'genealogical analysis', distinguishing it from his earlier archaeological analysis, where power was a less central concept (Foucault 1984: 133; Sheridan 1980: 116).

The result is a richly developed set of analyses, exploring how discourses constitute their objects and the rules of discourse formation. Such rules encompass the 'surfaces of emergence' (the social and cultural areas where a discourse first appears), the 'authorities of delimitation' (the institutional bodies that recognize and authorize a discourse), and 'grids of specification' (the various classification systems used). He also examined issues such as the status of speakers, the sites from which statements are made, and the positions of the subjects of discourse in relation to each other (Sheridan 1980: 96–9). Another aspect considered is the conditions governing the production of statements, culminating in the concept of the archive: 'the very system that makes the emergence of statements possible' (ibid. 102).

Foucault's interest in discourse/knowledge/power arose out of a more general reformulation of how the analysis of society should be approached. This is a radically decentred reformulation, in which the idea of the autonomous individual has no place. Rather individuals are defined, and even created by the articulation of societal forces. In earlier accounts, Foucault saw individuals as being defined by epistemes, a set of structural relations between concepts that defines their content. Concepts were specified according to structural forces, and they determined what individuals were, thought, and believed, and how they acted. Later, Foucault sought to move away from such an overly structuralist account and recast the creation of concepts in terms of discourses, seen as much more elastic or fluid. This further developed into an account in which discourses are seen as unstable, comprising multiple, even infinite, meanings (Bevir 1999). Such discourses do not reflect or map onto the position of dominant groups in society; just as power is a relational concept, so also is discourse. The shift from episteme to discourse, and in the view of discourses from stable to unstable, is associated with the shift in method from archaeology to genealogy.

The key consequence of such an approach is that the individual subject cannot be an agent of change. Change arises, instead, from the 'will to power', which is endlessly repressive; even as it constructs individual subjects, it does so only to repress them. So power both produces and represses, but in a decentred manner independent of individual action. Power is existent throughout society, often masquerading as knowledge or truth. This is also given expression in the form of institutions, seen as created and reproduced through the meanings of multiple, more or less random, micro-practices interacting. These processes by which institutions arise and continue to exist are contingent and, therefore, can only be understood through detailed study over time. And the same applies to government or the state, which operates in a similarly decentred way.

One important facet of social life that Foucault draws attention to is the way in which norms, laws, and rules are internalized, even to the extent of defining people's identities; he terms this pastoral power. But while many would recognize this as a feature of everyday practice and institutional life, it is more difficult to accept the limitations on the role of the individual that Foucault would attach to this. For pastoral power leaves little scope for the creative capacities of individuals, emphasizing instead the constraints of the regime of power. Furthermore, it becomes very difficult to conceive of change occurring other than in a random way; certainly it cannot be the product of intention and preconceived action. This is the main criticism that is levelled at Foucauldians. As Bevir (1999: 357–8) puts it: 'if the subject is a product of a regime of power, how can he act innovatively, and if he cannot act innovatively, how can we explain changes within a regime of power?' The only clue that Foucault offers to this problem is the possibility of resistance to the illusory freedoms imposed by the construction of the subject. Such resistance arises from the existence of power; it is power's necessary corollary.

Given this stance, it can be seen that the question of how actors use language is not a Foucauldian concern. This raises the question of whether a Foucauldian development of the social-constructivist paradigm is appropriate for the central concern of this study. Certainly, the very concepts of 'conflict', 'consensus', and 'language use', which have been used to frame this book, are at odds with the decentred nature of Foucault's approach to power. Conflict between actors (or consensus between them) is not the relevant unit of analysis for a Foucauldian. This, in turn, raises the question of whether a Foucauldian approach is necessarily structuralist. In a key commentary, Sheridan (1980: 90) argues strongly that Foucault's work is profoundly anti-structuralist. He emphasizes instead the way in which Foucault analyses the nature of historical change, and sees language as act or event, rather than structure. Discourse—understood as multiple, complex, and unstable—'transmits, produces and reinforces power, it also undermines, exposes and even blocks it' (ibid. 1986). While this concept of discourse as decentred action may or may not be structuralist (depending on how the term is defined), there remains little scope for the actor's own agency within a Foucauldian approach, and, for some, this is close enough to structuralism. As Darier (1999: 13) puts it: 'Foucault may have tried to "turn his back" on structuralism, but structuralism remains stuck to his back!'

This problematic can be made clearer by focusing on recent analyses of environmental planning that have sought to use the Foucauldian framework (such as Matthews 1996; T. Richardson 1996; Neale 1997; Sharp 1999; Darier 1999). These applications tend to work by taking concepts from Foucault's work and considering them in relation to environmental issues. This can be a difficult task, for Foucault was no Green or, to put it more elegantly (Levy 1999: 204): 'it would be a mistake to see Foucault or his poststructuralist colleagues as lending aid and comfort to the deep ecological project'. One example is the attempt to develop ideas of green or eco-governmentality from Foucault's conceptualization of the state (Luke 1999). This attempt is deeply problematic.

First, it presupposes a concern on the part of the state (usually conceptualized in unitary terms) in the environment that is on a par with the historic involvement of the state in the three great systems of exclusion outlined above. For example, Luke (ibid. 127) makes claims for the importance of a green governmentality based on the US government's environment policies, but is the environment really a central concern for the US government? Second, there is a tension between the tenor of Foucault's approach, which sees the state (and all other institutions) in repressive terms as a collection of instances of the influence of power, and the hope of most environmentalists (and environmental planners) in a potentially progressive environmental policy agenda. This can lead in two directions. On the one hand, it can result in an overly optimistic view of state action with regard to the environment, at odds with a traditional Foucauldian view of both the state and action. On the other hand, it can be used to support a view of the ecological agenda as itself repressive: 'by taking environmentalistic [*sic*] agendas into the heart of state policy, one finds the ultimate meaning of the police state fulfilled' (ibid. 149).

As Darier (1999: 27) himself points out, a Foucauldian approach is most helpful where either 'understanding the construction/deconstruction of subjectivities' is involved, or the 'practical political tactics of resistance' are being explored. It cannot be a primary resource in understanding the process of environmental planning as a whole. There are, though, two highly regarded analyses of environmental policy from a Foucauldian perspective, by Hajer (1995) and Flyvbjerg (1998*a*); how do they handle these problems?

Hajer's work is a study of acid-rain politics in Britain and the Netherlands. He uses the Foucauldian framework to consider how patterns of meaning surrounding environmental policy are related to structures of interest. In doing this, he finds that he has to modify and amplify the Foucauldian framework. First, he has to provide a role not just for the constraining aspects of discourse but also the enabling aspects (Hajer 1995: 49). Second, he recognizes the importance of interpersonal interaction when considering the micropolitics of environmental policy (ibid. 51). Third, he sees the need to develop 'middle-range concepts' to complete his task. In particular, he uses the concept of 'storylines' to develop a narrative analysis of policy discourses and he coins the concept of 'discourse coalitions' to identify groups of actors who share storylines, their storylines, and their discursive practices (ibid. 52–68). Members of a discourse coalition are, as a direct result of their membership, in a more powerful position.

These variations take the Foucauldian approach in a new direction. The question is whether it takes it so far as to represent a break point. Hajer does still hold to certain key elements of Foucault's framework, particularly its radical subjectivity with 'interests intersubjectively constituted through discourse' (ibid. 59), though this rather raises the question of whether actors have any material interest in environmental protection or sustainability. He also adopts the Foucauldian focus on hegemonic tendencies, which in Hajer's work is represented by the discourse of ecological modernization. Here he claims that ecological modernization has been a uniquely successful discourse of sustainable development, achieving hegemonic proportions; this is a view that some proponents of ecological modernization might find questionable.

Leaving these points of contention aside, all three of the modifications that Hajer makes do bring into question the Foucauldian view on actors' agency. If discourses enable, then what are they enabling? Surely this must mean that actors are enabled in their activities. If interpersonal interaction is important, then agency is important. And if discursive practices play a role in the operation of discourse coalitions, then discursive agency more generally is significant. This is a major disagreement with the essential elements of a Foucauldian perspective. Compare these two quotes, the first from Foucault (1984: 109), the second from Pocock (1984: 40), a political scientist interested in language and politics:

> in every society the production of discourses is at once controlled, selected, organised and redistributed by a certain number of procedures whose role is to ward off its powers and dangers, to gain mastery over its chance events, to evade its ponderous, formidable materiality.

> we act in ways consonant with language and yet unexpected; we reverse roles; we discover contradictions and negations; we set off resonances whose subversive tremors may be felt at the heart of the system; and we discover roles for ourselves in the teeth of the roles which language seeks to impose.

The Foucauldian approach has no place for the 'we' of Pocock's quote (see also Bevir 1999: 357–9). This has led many interested in discourse and attracted by the centrality of the concept in Foucault's work, ultimately to reject his approach. As Dryzek (1997: 20) says: 'Discourses are powerful, but they are not impenetrable (as Foucault and his readers have themselves inadvertently demonstrated in their own exposé of the history of various discourses; they at least have escaped from the prison of particular discourses!)'

A slightly different problem arises in Flyvbjerg's study of environmental planning in the Danish city of Aalborg, from an implicitly Foucauldian perspective (1998*a*; his espousal of this perspective is developed more explicitly in 1998*b*). The study concerns the planning of transport infrastructure in the centre of Aalborg over a period of some two decades. Interwoven into this story are details of pedestrianization schemes, the location of the bus station, connections to public transport networks, and road improvement plans. The plotline concerns the arguments for a more environmentally focused transport strategy and the way in which these arguments are overtaken by pressures for a more conventional, road-traffic centred mode of urban planning. Throughout the book, Flyvbjerg sets up a tension between rationality (the better argument) and power, his view being that power defines what is rational rather than rational arguments being an effective way of curtailing the exercise of power. This fits within Foucault's view of power masquerading as knowledge/truth/rationality.

The Foucauldian heart of Flvbjerg's (1998*a*: 25) analysis lies in his view of power. He sees everything that happens as evidence of or an expression of a power relation. Actual policy choices are, therefore, also expressive of power, and the form in which they are justified is a rationalization of that power relation and not evidence of rationality in making policy choices. The knowledge claims that support such rationalizations are also tied up with power relations, so that power defines what comprises knowledge in any situation. In the Aalborg case, the key actor is the business community. Its power rests in its political and organizational dominance (ibid. 30) and in its very focused and clear self-knowledge of its interests: it knows what it wants (ibid. 61). Furthermore, business is able to make use of a particular rhetorical strategy to generate a discourse that supports and generalizes its position. This strategy is a set of syllogisms whereby what is good for business is seen as being good for Aalborg, what is good for motorists is seen as being good for business and, therefore, what is good for motorists is seen as being good for Aalborg (ibid. 58). The role of the local newspaper (a member of the local chamber of commerce) is particularly important in promoting this discourse.

This is an intriguing analysis; many aspects of this story of rationalization within urban planning will resonate with those who have been involved in planning practice. It manages to connect issues of power, knowledge claims, and discourse within a coherent framework. And there is a quality of analysis that results from the detailed methodology of narrative and ethnography that Flyvbjerg employs. However, there are limitations, principally associated with the way in which power is theorized but also with the concept of rationality that is employed. The key problem is that all outcomes are seen as evidence of power relations. This means that there is no possibility of outcomes arising from any other source than the exercise of power. Dowding (1991) refers to this as the fallacy of attribution and points out that it ignores the possibility of actors benefiting from events over which they had no control or as a side-effect of someone else's decision-making. In a Foucauldian analysis, business benefits and, therefore, must be powerful. The current pattern of outcomes is seen as a consequence of power as well as reproducing power relations. As one might expect from a Foucauldian approach, this is essentially a structuralist perspective.

Within this, it is difficult to identify how less powerful interests might exercise any influence. Flyvbjerg locates the hope of such influence (admittedly a rather distant hope) in the exercise of rationality in conditions of stability. He points out that: 'In a democratic society, rational argument is one of the few forms of power the powerless still possess' (1998*a*: 229). The weakness of the conceptualization of power is caught in this quote. This provides very little practical scope for the strategies that weaker groups can adopt. In his discussion of Foucault, Flyvbjerg (1998*b*: 228) falls back on the politics of resistance and argues that 'Understanding how power works is the first prerequisite for action, because action is the exercise of power.' It is not clear how this statement relates to Foucault's radically decentred notion of the subject or to his systemic view of the pervasive operation of power.

Flyvbjerg has not been able to find a concept of power that is relational and not structuralist. Dowding (1991), in seeking to avoid the fallacy of attribution, turns towards an individualist notion of power as something that actor A exercises over actor B. However, Flyvbjerg follows Foucault in seeing power as relational, where power vests in the relationships between people. The problem with such a relational concept of power is that it becomes difficult to specify the relative degree of oppression involved at any particular moment. If power is a set of unequal shifting relations (Sheridan 1980: 184), how does one determine who is being favoured? This requires an additional framework of understanding for the networks of relations that exist between actors. Without this, and with an emphasis on empirical outcomes as the test of power, the analysis inevitably tends to the structuralist.

The other problem is that rationality becomes an almost meaningless concept within this analysis, not least because of the multiple wordplays on rationality and rationalization that Flyvbjerg employs. Rationality becomes a synonym for justification, without any autonomous meaning. And although Flyvbjerg tries to recoup some independent meaning by identifying the very limited circumstances in which rational argument may come to the aid of less powerful groups, this is very difficult within a framework that explicitly states that reality is defined by power. What impact can rationality have on reality if power is so all-pervasive as to define it? What is needed here is a better understanding of how rationalization works as a discursive strategy, so that it can be delinked from the notion of rationality as the better argument. This would involve more detailed attention to the language of rationalization.

This discussion of Foucauldian applications to environmental planning issues has highlighted key limitations concerning the role of agency, the view of the state as an institution, and the conceptualization of power that is involved. To explore an alternative contemporary paradigm, the Habermasian framework will now be addressed.

The Promise of Communicative Action: Ecological Democracy and Collaborative Planning

The very title of Habermas's *Theory of Communicative Action* (1984, 1987) indicates the central role ascribed to agency. Habermas's framework has also proved a fruitful starting point for those seeking to develop normative theories of planning practice or models of democratic practice oriented towards ecological sustainability. Communicative action refers to action oriented towards mutual understanding through the use of language; successful communicative action can play a central role in achieving social co-ordination. As a concept, it recommends an examination of the structures that enable competent speakers to engage in successful interaction. It is counter-posed to strategic action, in which actors pursue their individual interests. Within communicative action, success is measured with reference to the level of mutual understanding that is achieved between actors. The concept of rationality itself is recast in this light: 'Communicative rationality is found to the degree that communicative action is free from coercion, deception, self-deception, strategizing, and manipulation' (Dryzek 2000: 22).

In Habermas's view, strategic action—with its emphasis on the purposive, the individual, and the interest-based—cannot be an adequate basis for understanding the legitimacy of contemporary society nor for developing a model of democracy. It is, therefore, a major concern that, within modern society, strategic action and instrumental rationality are becoming dominant. They are seen to be moving outwards from their societal base within the economy and the political system and into the 'lifeworld', the arena of everyday interactions between individuals within families, households, and personal lives where communicative rationality has traditionally held sway. To counter this, Habermas proposes that we should build on the fragments of communicative reason that already exist within the lifeworld to create an alternative deliberative form of democracy.

The hope of communicative rationality lies in the potential of language, not just to support the exchange of statements but actively to foster dialogue between actors. For Habermas, communication necessarily carries with it the promise of moves towards mutual understanding—and even agreement—through a process of reasoned argument. Because of language, consensus is not an impossible ideal but already nascent within the communicative situation. However, this remains only a promise, an immanent possibility, which cannot readily be made an explicit and acknowledged element of political life. For the more negative message of Habermas's approach is that we live in a world of distorted communication, in which powerful interests shape forums and modes of communication systematically to advantage and disadvantage different groups. Through this idea of distorted communication, the analysis of communication between actors is systematically linked to that of power and interests. But communication cannot be reduced to the play of interests; contemporary communicative structures are given an independent dynamic influence of their own.

This kernel of Habermas's work is at the heart of two rather different applications relevant to environmental planning: Dryzek's extension of deliberative democracy into discursive and ecological democracy and Healey's development of the normative planning theory of collaborative planning. Both identify the potential of the Habermasian framework but also its limitations as a basis for understanding contemporary environmental planning practice and the role that language can play. This discussion is extended by considering

Mason's application of Habermas's and Dryzek's ideas to a variety of cases of environmental policy.

Dryzek's work is closest to Habermas's in terms of the abstract level of analysis and links to political and social theory. Like Habermas, he is concerned to develop the normative dimensions of an alternative theory of democracy. However, as someone with a sustained interest in environmental issues, he is also concerned to build a model of political processes that has a distinctively ecological dimension. His focus has, therefore, been to work out the details of a programme for reshaping democratic practices from this perspective. His aim is 'collective decision-making through authentic democratic discussion, open to all interests, under which political power, money, and strategizing do not determine outcomes' (Dryzek 1990: 29–56; 1997: 199).

For Dryzek, a democratic society is one where communicative rationality prevails and where all impediments to that distorted communication are overcome. If this situation is achieved then Dryzek (1997: 200) believes that certain values will also prevail: 'the kinds of values that survive authentic democratic debate are those oriented to the interests of the community as a whole, rather than selfish interests within the community (or outside it)'. This would include green values, because communication would extend to the non-human world: 'the nonhuman world can communicate, and human decision processes can be structured so as to listen to its communications more or less well' (ibid.). Dryzek (1995: 20) describes this as trying 'to rescue communicative rationality from Habermas's by extending it non-anthropocentrically. He terms this normative ideal 'ecological democracy', itself an extension of the more general notion of 'discursive democracy'. In working out this programme, Dryzek (2000: 2) remains aware (though not always consistently) of the need to maintain a credible stance in relation to contemporary power structures. Indeed he distinguishes ecological and discursive democracy from the more general concept of 'deliberative democracy' precisely in order to retain a more critical stance.

However, even with a more overtly critical element built into his theorizing, Dryzek's work remains primarily a normative programme. It does not, on its own, provide an analytic framework for understanding the current role of language in relation to conflict and consensus in environmental planning. This is because the transfer of concepts from the realm of political theory to policy practice remains problematic. Communicative rationality is oriented towards communicative, mutual understanding; it is opposed to goal-oriented behaviour, the realm of instrumental rationality. As such, communicative rationality can form the main or sole basis of a normative programme and, in a broad sense, democracy may be seen as a process of mutual comprehension. But as a concept for analysing contemporary practice, it is far too limited. In most policy arenas, mutual understanding is an insufficient goal; some other essentially instrumental goal is at issue, whether it is deciding on a proposal or negotiating an agreement. Parties to the policy process rarely feel that mutual understanding on its own is a sufficient reward for investing their time and effort. So instrumental goals tend to enter into or even form the starting point of the process, compromising the reliance on communicative rationality.

This tension between the normative and empirical, between ideals of democratic practice and current patterns of social action, between thought experiment and critical analysis is clearly identified in Habermas's own work, particularly in *Between Norms and Facts* (1996). Here Habermas makes quite clear the assumptions and limitations that constrain his own model of deliberative democracy. In particular, he emphasizes how the exercise of communicative action in order to reach mutual understandings has to be grounded in shared backgrounds, which support shared expectations, resources, and identities. In a complex modern society, multiple lifestyles and social groupings undermine the potential for successful communicative action. He points out that conflicts can only be resolved on the basis of reasoned argumentation if:

- members of a community mean the same thing by the same words and expressions;
- members consider themselves mutually accountable and ready to take on obligations arising from agreements; and
- arguments supporting a mutually acceptable resolution are not found to be false or mistaken.

This means that disagreements on some matters will be much more difficult to achieve than on others, depending on whether these conditions are fulfilled or not. It also points to the need for a fuller institutional analysis of the bases for communication.

More fundamentally, it is only possible to replace strategic by communicative action with the 'performative attitude of a speaker who wants to *reach an understanding*' (ibid. 18; the emphasis is in the original, although I would prefer to emphasize the 'want'). Where agreement cannot be reached, Habermas identifies only a limited number of options, the most likely of which is shifting into strategic action, and bargaining to resolve conflicts (ibid. 21). For these reasons, he states that communicatively achieved agreements are always open to challenge and are a precarious source of social integration: 'the ever-present risk of disagreement built into the mechanism of reaching understanding' (ibid.). The point is that Habermas himself recognizes the limited application of his framework for contemporary critical analysis and also for changing policy practice. This was not his primary intention in a work of political philosophy: 'communicative rationality is not an immediate source of prescriptions' (ibid. 4).

This point is, perhaps, not sufficiently appreciated within Healey's *Collaborative Planning*, a normative theory of how local environmental planning should operate, which was developed in response to some three decades of debate within planning theory, during which faith in a rational decision-making form of planning theory was undermined by a number of factors.

These include: an internal critique, the evidence of numerous planning failures, and the emergence of a political economy analysis emphasizing the inequalities and power games inherent in planning processes and outcomes (Rydin 1998*c*: ch. 2). Healey herself undertook a number of research studies that used a modified political economy approach but, rather than rest the analysis there, she sought to find a renewed role for planning that was less naïve about its processes and consequences than traditional rational-comprehensive or procedural planning theory. The interpretative and communicative turn in social sciences and Habermas's theory of communicative action provide the launching point for such a normative account, which recasts planning as 'a process of interactive collective reasoning, carried out in the medium of language, in discourse' (Healey 1997: 53).

The balance and, importantly, the style of her work tend to emphasize the programmatic aspects of the Habermasian approach at the expense of the critical. It is not that she ignores issues of power; her main text explicitly discusses power relations and how they can be conceptualized (ibid. 112–19). But, having conceptualized power in terms of relational webs, she emphasizes the potential for changing these relations and the role that the planner can play in bringing change about (ibid. 118–19): 'Social change may therefore be encouraged by strategies which aim at opening up relational links and challenging exclusionary ones where these reinforce inequality. . . . A recognition of power relations of everyday life experience is of critical importance in development of the practices for collaborative local planning.' Here the processes of planning (and, by implication, the role of the planner) are seen in terms of transforming the public realm, changing structural forces, and having a 'creative capacity, to help build new "transcending principles" and practices, which change the systems through changing the cultural reference points of all those involved' (ibid. 55). Healey remains convinced that communication can result in consensus in practice: 'If the culture-building process of strategy-making has been rich enough and inclusive enough, the strategy should have become widely shared and owned by the participants and the stakeholders to whom they are linked' (ibid. 279).

The emphasis on potential rather than constraint arises because she sees differences between actors and groups not so much in terms of interests but 'deeper, in ways of being, of giving meaning and value to things and relations, and in styles of expression' (ibid.). Power is thus revealed through communication and reflection. This effectively reduces interests to discourse but, unlike the radical Foucauldian approach, then renders interests capable of change through communication and reflection. This in turn reduces power to discursive acts (not to discourse, as in a Foucauldian approach). It is telling that the section on power relations is mainly to be found in the chapter on 'Everyday life and local environments' rather than in the chapter on economic processes, where a rather different story might need to be told. Healey (ibid. 66) states that 'We cannot ignore that we live in a world which is heavily structured by powerful forces,' but she then goes on largely to talk about how this can be overcome.

Consequently the criticism has been levelled at Healey that she is effectively wishing away the issue of power (Flyvbjerg 1998*a*; Tewdwr-Jones and Allmendinger 1998). This apparent neglect arises from her concern to adopt a model of individuals as reflective beings rather than subject to structuralist forces (Healey 1997: 57) and is also related to her adoption of the metaphor of 'networks' rather than 'structures', again suggesting more scope for change and fluidity. The balance between structural constraint and change is identified but the discussion often tips in favour of the potential for change: 'These relational encounters over shared local environmental issues reflect power relations. But the potential always exists, however small, to transform them' (ibid. 60–1). This contrasts with the view of Forester (1989: 100), the American planning analyst, who argues that planners 'have to be clear about what mediated negotiation [between parties in the planning process] will *not* do: it will not solve problems of radically unbalanced power, for example'.

Phelps and Tewdwr-Jones (2000) further develop the critique of collaborative planning by pointing out that Healey emphasizes only one aspect of social action identified by Habermas. They point to three other categories of social action: teleological action, which is goal-oriented means-end action; normatively regulated action, which is the use of predefined common values to ensure individuals comply with the group's norms; and dramaturgical action, which concerns the presentation of the self to a specified audience. They then argue that these different types are all inherent in communicative action and do not operate distinctly. Any example of practice will spread across these categories, thus rendering problematic the claims for a specifically collaborative planning based solely in communicative action, and suggesting the need for an approach that encompasses communication and strategic dimensions to policy action.

This is interesting because it highlights the rather limited view of language, discourse and communication that is inherent in collaborative planning. For Healey's view of discourse is of an 'argumentative jumble' (1997: 275), loosely structured and amenable to change: 'Through the development of ideas in *policy discourse*, systems of meaning can be changed' (ibid. 61). It does not consider how this can happen or when it will fail. Neither does it consider how discursive systems may be structured to give rise to different types of social action, for dramaturgical and normatively regulated action are primarily discursive and even teleological action has a discursive dimension. Nor does it suggest how discursive systems may constrain action, even communicative action. As Burgess, Harrison, and Filius (1998: 1447) sum up: 'Healey perhaps fails to acknowledge sufficiently the differences in discursive power between those with technical, professional expertise, and lay people. There is a presumption, as with Habermas's original formulation, that with sufficient "goodwill" and sensitivity to difference, that more equitable representations of different cultures will be achieved in policymaking.'

Despite these criticisms, the great strength of Healey's work has been the way in which it has brought attention to and supported both theoretical debate and substantial empirical work on the communicative dimensions of planning practice. This work has not been limited to specific stages of the policy process but has ranged across many different dimensions of the planning process. It has given many insights on the detail of planning practice, which will be drawn upon later in the book. However, it remains questionable whether the analytic framework presented in *Collaborative Planning* itself is sufficiently robust to support the full understanding of the role of language within planning processes. The normative emphasis turns attention too quickly away from the critical edge in Habermas's work, which itself could provide the basis for a more analytic and critical framework. As Dryzek (noted above) and other authors have recognized, there is an explicit need to maintain a critical edge.

Mason has sought to maintain that edge by applying a 'discourse principle' derived from the Habermasian framework in both a normative and explanatory manner during his analysis of environmental policy and planning. The normative dimension revolves around 'a radical democratic project which extends and radicalizes existing liberal norms in order to include the ecological and social conditions for civic self-determination' (Mason 1999: 9). This is closely aligned with Dryzek's ecological democracy, although Mason rejects the possibility of non-human nature qualifying to participate in democratic dialogue, favouring instead a humanistic perspective (ibid. 56, 62). However, Mason also seeks to use the discourse principle to account for 'existing tendencies for non-coercive green communication' (ibid. 9), in an attempt to use Habermas's framework to do explanatory work. This works largely by judging the 'quality of democratic communication on environmental issues' (ibid. 15). It is questionable whether this actually extends normative judgement into explanation. By applying the same standard to develop the concept of desirable political practice and to assess contemporary practice, the normative is not balanced by the explanatory but rather the scope of normative judgement is extended from the hypothetical ideal to current activities. This may be worthwhile but it is not explanatory analysis. Thus, when Mason identifies a gap between public opinion and that of environmental NGOs, he argues that these organizations still have 'communicative work to do' (ibid. 89), that is, that they need to engage in persuasion of public opinion. This is a normative stance, rather than an explanatory one.

The empirical material that Mason presents actually highlights the severe problems that are encountered in moving towards the ideal of environmental democracy. These include:

- considerable practical problems of time, resources, expertise, and political will;
- not all parties being willing to participate;
- failure to achieve full consensus;

- the consensus reached not necessarily being benign;
- groups often engaging in strategic lobbying and acting to protect their interests, rather than trying to reach mutual understanding;
- key stakeholders being unwilling to cede control; and
- problems of lack of accountability.

He concludes that examples of democratic practice were isolated and often unlinked to other levels of government. These are an important set of findings from the case study research. But Mason's use of the discourse principle means that he looks for the 'silver lining to the cloud', the aspects of democratic communication that can be identified, rather than asking: why are there these problems with attempts at environmental democracy? What are the key factors that may help explain these patterns? To answer such questions, it is necessary to go beyond the Habermasian framework.

Conclusion

This review of various approaches from within sociology, political science, and environmental planning studies has highlighted the benefits of 'revealing the hidden' and conclusively demonstrates the need to take into account how issues are constructed, how agendas are built, and how broader discourses within society influence policy processes. In particular, it has pointed to the importance of discourses of rationality within legitimization processes. However, the review has also highlighted problems with existing approaches. Many approaches, both within political science, sociology, and, specifically, Foucauldian frameworks tend towards the structuralist and do not allow a sufficient space to explore the role of agency and actors' strategies. Many of the approaches discussed pay inadequate attention to the detail of how discourse works, how language is used, and actors' discursive strategies. While the Habermasian framework is a fruitful basis for normative theorizing it does not appear to have provided the best approach for critical analysis of the policy process, including environmental planning. It remains a primarily normative framework. There is a clear need to go beyond the issues raised by this framework, to develop an explanatory model of the policy process in which the role of language in handling conflicts, generating consensus, and legitimizing policy can be fully explored. This is the task of the next chapter.

3

Discourses, Communication, and Discursive Strategies: An Institutionalist Framework

The previous chapter has highlighted two very different dominant perspectives on the analysis of discourse within the environmental policy process. Social constructivist and Foucauldian approaches suggest a deconstruction of environmental discourses to identify their impact in terms of knowledge claims and power relations. More actor-centred approaches highlight the role that purposive action can play, and the Habermasian framework suggests a normative project for democracy in which communication through discursive acts could lead to mutual understanding, consensus, and greater legitimacy. None of these approaches was seen as ideal for analysing contemporary practice of environmental planning and specifying the role that language actually plays. It was identified that a more agency-oriented perspective was required than that provided by Foucauldian approaches, and a more critical as opposed to normative stance than that provided by Habermasian approaches. In either case, a more detailed attention to language and how it is used within the policy process is needed.

This chapter, therefore, sets out a framework that emphasizes the level of planning practice and seeks to allow for both the constraints placed on that practice by the influence of discursive patterns and the potential for achieving change through discursive strategies. In doing so, it is the concept of the 'institution' that deserves more attention. The discussion begins with this concept, placing it within the general trend towards institutionalism and outlining the approach adopted by the Ostroms, called institutional analysis and development (IAD). Particular attention is paid to the model of agency and of actors that is implied by this framework. The ways in which a concern with language can be explicitly integrated into this approach are then considered. The framework proposed adopts a three-dimensional approach encompassing discourses as cultural patterns, the practice of communication, and purposive discursive strategies. This helps explain the formation and influence of discourses and the active operation of institutions.

Institutions: The Key Focus

The concept of institutions has already been mentioned in several contexts in Ch. 2. For Foucauldians, society is defined by major institutions, characterized and maintained by certain discourses and knowledge/power patterns. In this view, institutions are large-scale societal entities. For Habermasians, there is more of an emphasis on institutions, which can be created and maintained as arenas for conducting interaction between social actors. Institutions are spaces where communication, dialogue, debates, and deliberations occur. Following the 'new' institutionalism, originally associated with March and Olsen (1989), institutions are defined in terms of norms and routines of working practice. Classifying both these aspects under the term 'rules', March and Olsen (ibid. 22) incorporate: 'the routines, procedures, conventions, roles, strategies, organizational forms, and technologies around which political activity is constructed. We also mean the beliefs, paradigms, codes, cultures, and knowledge that surround, support, elaborate, and contradict those roles and routines'.

Such rules can encompass many different aspects of practice: procedures, decision-making, evaluation, allocation of authority and responsibilities, record-keeping and information-handling, access, rights of opposition, timing, and even changing the rules themselves. Such rules may be consciously designed and clearly specified, or they can be unwritten customs and conventions. And they operate at many different levels, from society-wide rules through to locality-specific and organizationally sited rules.

This emphasis on the institution as a routinized set of working practices and everyday operational activities associated with norms and values is conceptually distinct from the organizational arrangements that actors have to operate within. For example, local authorities are particular organizations, structured into departments, committees, or other units. But these organizational structures do not determine working practices. Any actor working within a local authority will find him- or herself subject to the prevailing norms of working practice and, indeed, may well face overlapping and even competing norms; these norms represent institutions. Thus a local authority environmental officer may be subject to the institutions associated with local government generally, with this particular local authority, and with her professional affiliation. Consider a case where local government is being 'modernized' along private sector lines with the emphasis on competition and market-based pricing, the specific local authority had a tradition of old-style labourist politics (and perhaps a history of corruption) and the profession is a technocratic and scientific profession organized along collegiate lines. Then the actor might be subject to considerable tensions between the norms, expectations, and routines suggested by these three institutions.

There can also be tensions between old and new institutions, existing patterns and innovations. Lowndes (1999: 24) points out that institutions are, by their very definition, associated with stability and slow rates of adaptation. But innovations do occur and new institutions are created. In this case, old and new institutions have to coexist and actors have to cope with both. Similarly, actors have to cope with institutions at different levels and the overlapping influence of multiple routines. As March and Olsen (1989: 22) point out, this means that it is not a simple matter to identify the influence of rules, even when behaviour is seen as primarily influenced by such rules. Inconsistencies between rules are common and so actors often have to choose between them. The claim of new institutionalism is that these rules are nevertheless the key to understanding organizational practice, such as environmental planning.

March and Olsen (ibid. 23) see rules as 'driving' behaviour, although they cast this in terms of actors' engaging in search and choice in the context of rules:

> To describe behavior as driven by rules is to see action as a matching of a situation to the demands of a position . . . Political actors associate specific actions with specific situations by rules of appropriateness. What is appropriate for a particular person in a particular situation is defined by political and social institutions and transmitted through socialization. Search involves an inquiry into the characteristics of a particular situation, and choice involves matching a situation with behavior that fits it.

Lowndes (1999: 23), using the new institutionalism to help understand the processes of local governance, similarly emphasizes how rules can embody a 'logic of appropriateness', that simplifies the choices facing actors and, in this way, can 'guide and constrain people's action'.

This involves an emphasis on obligatory action. March and Olsen characterize obligatory action in terms of the following questions, which actors pose to themselves:

- What kind of situation is this?
- Who am I?
- How appropriate are different actions for me in this situation?

And they contrast this with the following questions, arising from an emphasis on anticipatory action:

- What are my alternatives?
- What are my values?
- What are the consequences of my alternatives for my values?

Such anticipatory action is more commonly associated with rational-choice accounts or with strategic action. The two types of action are seen by March and Olsen as mutually exclusive, with more explanatory power residing in obligatory action and, hence, in institutional norms and rules.

Since rules are reproduced through actors' search and choice of appropriate behaviour, rules persist over time and can become the embodiment of power. Focusing on rules can lead to an understanding of how actors, who have already acquired resources and power, are able to use rules to their benefit. Lowndes, for example, identifies how rules can justify particular distributional outcomes, and March and Olsen (1989: 47) specifically state that 'There is a tendency for large, powerful actors to be able to specify their environments, thus forcing other actors to adapt to them. Dominant groups create environments to which others must respond, without themselves attending to the others'. However, there remains the scope to create new institutions, which try to redress this imbalance. As Lowndes (1999: 23) points out, this may not be an easy task as it involves both 'strategic action to create new rules and incentives', and also 'norm-governed behaviour to embed and sustain new rules over time'.

This emphasis on rule-driven behaviour can lead to a rather functionalist account. Indeed March and Olsen make claims for rules in terms of the functional role they can play in society. They argue that rules enable a number of different facets of social interaction, namely: co-ordination, conflict avoidance, and interpretation. They also encourage realistic bargaining by constraining bargaining activities to the feasible and they mitigate uncertainty by their very existence (ibid. 24). In all these ways the simplifying effects of rules are beneficial to societal maintenance. However, this rather functionalist approach is not a necessary corollary of a focus on institutions. It is possible to accept that institutions are an important focus of analysis while recognizing that institutions may fail to perform the integrating and simplifying functions that March and Olson envisage, or that they do so in a perverse and dysfunctional manner.

Similarly, the exclusive emphasis on obligatory action that March and Olson espouse is unnecessarily restrictive (and can be likened to the overemphasis on communicative action at the expense of other forms of social action noted in Ch. 2). Anticipatory and obligatory action can and frequently do coexist in contemporary social practice. Ostrom has developed an approach that allows for both norm-regulated, rule-driven behaviour and also an anticipatory, cost-benefit calculus on the part of actors. In specific situations, one form will dominate but over time both forms will be used. And, indeed, it is not always that easy to separate them. Any anticipatory calculus takes place in the context of particular sets of rules, while routine following of rules may be based on an implicit calculus that this is the easiest, quickest, and least effortful pattern of behaviour to follow. In E. Ostrom's account, both these kinds of interconnection between obligatory and anticipatory action can be found, although most of the emphasis is on the former—the role of rules as contexts for decision-making. This slight shift in emphasis is reflected in her preferred definition for this central term 'institution'. Ostrom (1999: 37) uses the term to refer to 'the shared concepts used by humans in repetitive situations organised by rules, norms, and strategies' and then uses this to frame an approach that she terms the institutional analysis and development (IAD) framework. To reiterate, the particular value of this framework lies in the way in which it manages to integrate a focus on norms with analysis of actors' interests seen in terms of the incentives (and disincentives) facing actors. It is summarized, in illustrative form, in Fig. 3.1.

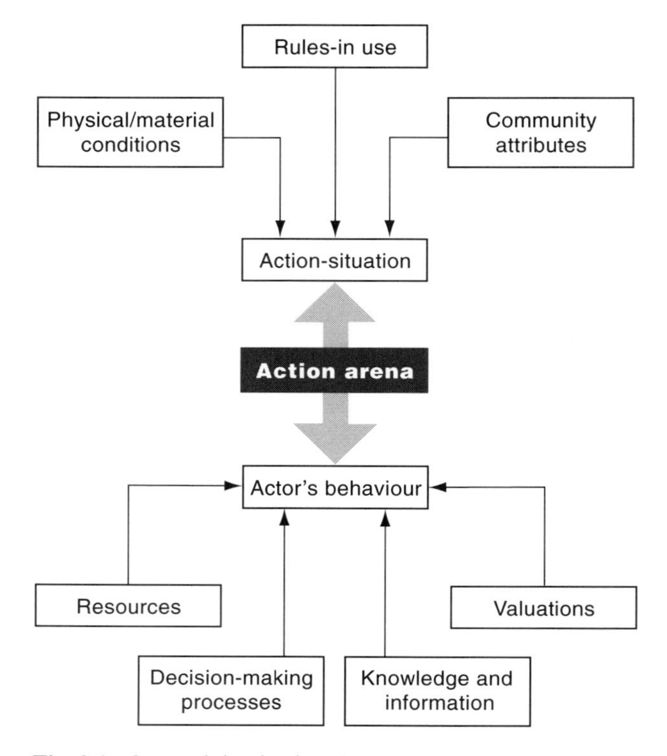

Fig. 3.1. Ostrom's institutional analysis and development framework

The central element of the IAD framework is the action arena, where behaviour and decision-making generate patterns of interaction and hence outcomes. This action arena is defined as 'the social space where individuals interact, exchange goods and services, solve problems, dominate one another, or fight', among other things (ibid. 42). It comprises two elements—the various specific action situations and the actors themselves—and is subject to three sets of influences: rules-in-use, the attributes of the particular community under discussion, and physical/material conditions. The relative importance of rules-in-use, material conditions, and community attributes will vary with the specific circumstances of the case. In some cases, material conditions are not particularly significant, while in others they are severely constraining on possible outcomes. Community attributes are more likely to be generally significant; these attributes include norms of acceptable behaviour, the level of common understanding, the extent of homogeneity, and the distribution of resources. But in all cases, IAD assumes that rules—shared understandings about what actions or states are required, prohibited, and permitted—are important determinants. Such rules will operate at several levels: operational rules affecting day-to-day decision-making; collective-choice rules determin-

Types of rule	Characterization of action situation
Entry and exit rules	Who participates
Position rules	Their positions
Scope rules	Possible outcomes
Authority rules	Action–outcome linkages
Aggregation rules	The control that participants exercise
Information rules	Information
Pay-off rules	Costs and benefits assigned to outcomes

Fig. 3.2. Rules and action situations within the IAD framework

ing change in operational rules; constitutional-choice rules determining change in collective-choice rules; and even metaconstitutional rules underlying all the others.

All aspects of the action situation can be specified by reference to the prevailing rules (see Fig. 3.2). This means that any particular situation in which decision-making is occurring can be characterized in terms of these variables and, more significantly, in terms of the rules shaping decision-making. Specify the rules and the nature of interaction within the action situation is partly specified. Where these rules persist over time and become routinized, they may become less overt and obvious but they will continue to influence decision-making and policy outcomes. It is important to note that the costs and benefits of outcomes (the incentive structure that actors face) is also specified in terms of rules, pay-off rules. The material and the rule-based nature of incentive structures facing actors are both recognized and seen as interrelated. Among the incentives that Ostrom (1992: 24–5) lists as significant within these structures are: material inducements, opportunities for prestige, desirable working conditions, pride, personal comfort and satisfaction, conformity, and the feeling of participating in important events.

The other important elements of an action arena are the actors themselves. However, a focus on actors does not preclude a concern with the group; actors can be individuals or corporate entities. Neither does it deny the importance of structural patterns, which in IAD are constituted by the three variables of material conditions, community attributes, and rules-in-use, each operating at successively higher levels. Feedback loops operate from the outcomes (generated by interactions between actors) to the variables, so that structural patterns influence but do not determine, are significant but not immutable. The dynamics of the framework come, though, from the actors 'giving life' to the contextual variables. As Fig. 3.1 indicates, actors are characterized by four features: resources; the valuations that actors assign to states of the world and actions; the ways in which actors acquire, process, retain, and use knowledge and information; and, finally, the processes that actors use for the selection of particular courses of action.

What such an institutional analysis seeks to achieve is: first, an understanding of the structure of the action arena (actors and situation interacting); and second, an explanation of this and the resultant outcomes in terms of the three key variables of rules-in-use, community attributes, and material conditions. Its strengths lie in the way in which agency is given a clear role but set within context. By using the concept of institutions to frame the analysis, rules-in-use provide the linkage between agency and context. Since such rules-in-use (and also various other elements of this framework) have to be expressed discursively, his framework provides an ideal 'docking point' for a discourse approach, that connects to interests, conflicts, consensus-building, and rationality claims. First though, it is appropriate briefly to consider the model of the actor implied by this approach.

As Ostrom makes clear, the way in which the interrelationships within the IAD framework are modelled depends on the assumptions that the analyst makes about elements of the action arena, including assumptions about the actors themselves. Ostrom herself is closely associated with the rational-choice school of political science. Rational choice is characterized by its adoption of a working model of the individual in the decision-taking situation, which emphasizes instrumental rationality, that is, that actors have goals and pursue them. In its most extreme or rigorous formulation (depending on your point of view), this instrumental rationality will be limited to material considerations and will be clearly specified in terms of ordered and self-consistent preferences. Actors will have utility functions that determine their decision-making behaviour. This is a highly reductionist view of actors and one that is empirically implausible. However, Ostrom makes it clear that this is not a necessary characterization of actors within the IAD framework.

What is required is a conceptualization of actors that allows for active decision-making but within more realistic parameters. Actors are both instrumental and psychologically complex, capable of drama as well as logical thought. They are also shaped by cultural factors, by the various institutions they are part of. This gives expression to their social nature. Some rational-choice adherents try to subsume the apparently non-instrumental into instrumental strategies for decision-making. Thus, if an urge towards altruism is a psychological trait, then it becomes rational for actors to behave altruistically; it is in their interests to do so because they gain psychic benefit from altruistic behaviour. As many critics of rational choice have noted (Green and Shapiro 1994), this approach runs the danger of descending into tautology. Any outcome is read back into an assumption of rational behaviour; if an actor behaves in a certain manner, this is because it is in her interests to do so and her interests have to be respecified until the benefits outweigh the costs.

A better approach is to say that we operate with a mixture of motives and interests, individual and social pressures, and the precise mixture depends on the situation under consideration. It is necessary, though, to make an assumption about the mixture that is likely—in most cases—to be operating in a given context. Thus, for most familial contexts a set of motives, which include love, jealousy, other emotions, and certain neurotic attachments may be considered to dominate. But in the policy context—which is the subject of this book—it is likely that the pursuit of material and social interests will be a dominant feature of actors' motives in most cases. The task becomes one of identifying how the processes of pursuing such interests work. Included here is the way that benefits and available strategies for self-advancement are perceived. There is no need to assume clear-sightedness on the part of individuals in relation either to their benefits or strategies. But neither is it realistic to assume a complete lack of appreciation of interests, benefits, and appropriate strategies.

The notion of rationality involved in actors' decision-making can, though, be recast in a more realistic direction. Allowances can be made for making mistakes and opportunism and actors can be reframed as 'fallible learners' (E. Ostrom 1999: 45–6). This is a form of 'bounded rationality'. Here, rationality is no longer the synoptic ideal of comprehensively considering all possibilities; rather it is a bounded concept, bounded—as Coyle (1997: 60–1) puts it—by culture and by language. Bounded rationality implies that actors are essentially reactive and will work along given lines until prompted to act otherwise, by an outside event or other change. But then they can and will act proactively. Bounded rationality, like institutions themselves, privileges simplicity and repetition, with proactive strategies arising only as needed (Ostrom, Gardner, and Walker 1994: 199). This enables the basis of rationality to be broadened to consider much more than the instrumental materialism of conventional rational choice. That is, it can incorporate norm-following and obligatory behaviour alongside anticipatory behaviour, other forms of social action alongside communicative action.

Institutions and Discourse

As currently formulated by E. Ostrom (1999: 51), the discursive dimension is largely missing from the IAD framework. She makes it clear that 'All rules are formulated in human language' and depend on 'the shared meanings assigned to words used to formulate a set of rules', and yet the role of language is not fully developed. This dependence on language is just seen as a potential pitfall, a source of ambiguity and misunderstanding: 'rules share the problems of lack of clarity, misunderstanding, and change that typify any language-based phenomenon' (ibid. and also 1994: 40). This is a very limited appreciation of the role of policy discourse. Similarly the founders of new institutionalism, March and Olsen (1989: 25) make only limited reference to language. They note that 'the process [of relating rules to situations] is heavily mediated by language' and they also point to several different dimensions of language use. For example, they look at policy and politics in terms of symbolic action (like Edelman 1984) and see such symbols as a form of strategic action itself (1989: 47–52). However, the discussion of how language works does not go much further.

Perhaps at this stage it is necessary to clarify exactly how discourse can be used as a term. There is considerable variety in the way the term is currently used. Milton (1993: 8) points out that discourse has meaning both as a process and as substance. As a process it refers to the way in which social reality is constituted by the organization of knowledge in communication; as substance, a discourse is a field of communication defined by its subject matter or type of language used. Hajer (1995: 44) manages to combine both elements in his definition: 'Discourse is here defined as a specific ensemble of ideas, concepts and categorizations that are produced, reproduced and transformed in a particular set of practices, and through which meaning is given to physical and social realities'.

But definitions of discourse can also vary in terms of whether they choose to emphasize the processes of discourse creation or the hermeneutic aspects of how they are understood. Dryzek (1997: 8) focuses on the later: 'a discourse is a shared way of apprehending the world'. More fully:

> Embedded in language, it enables those who subscribe to it to interpret bits of information and put them together into coherent stories or accounts. Each discourse rests on assumptions, judgements, and contentions that provide the basic terms for analysis, debates, agreements, and disagreements, in the environmental area no less than elsewhere. Indeed, if such shared terms did not exist, it would be hard to imagine problem-solving in this area at all, as we would have to return to first principles continually.

Seeking to provide a classification of discourse definitions, Van Dijk (1996) identifies four alternative uses. Discourse can be used to mean language use, the communication of beliefs, more generally as interaction in social situations, and finally, in the Foucauldian sense, knowledge and power.

Rather than having to choose between these different uses, a fuller picture of how the policy process works discursively can be obtained by looking at three different dimensions to discourse and policy, three different discursive 'slices' through the policy process. These three 'slices' relate to different levels of analysis and different dimensions of policy action. They move from the most general level through a concern with arenas down to a more actor-centred concept. In this way, the analysis ranges across many of the dimensions identified by the above authors as relevant to a concern with discourse and language. In the following sections, these three dimensions will be explored further. First, the ways in which prevailing patterns of language use structure and constrain the options open to actors are explored. The emphasis here is on the regular patterns that manifest themselves in language use, and the term 'discourse' is hereafter reserved for these patterns and patterning processes. Second, there is an emphasis on the way in which the policy process opens up (or closes down) opportunities for communication and the way in which that communication is structured, including the organizational and normative aspects of communication. And third, there is the active use of language in discursive strategies by actors. All three dimensions are relevant to an institutionalist approach.

Institutions and Patterns of Discourses

Dryzek's (2000: 18) use of the term 'discourse' relates to the patterns arising from language use. He describes discourses as 'institutional software'. This definition also fits within Milton's (1996: 166–7) specific level of discourse as 'a field characterized by its own linguistic conventions' as opposed to a general process through which knowledge is constituted or a specific area of communication defined by its subject matter. Such patterned forms of language have a much richer influence on policy than simply the ambiguity highlighted by Ostrom. Looking at the seven types of rule identified within the IAD framework in Fig. 3.2, the operation of all these will be influenced by the ways in which language works, the opportunities it offers and those it denies in the particular context. So if a rule is couched in terms that allow a certain range of interpretations, this will be a different situation from an even slightly different wording, allowing a different interpretation. Furthermore, the rich range of rhetorical tropes available will affect the use of language, regardless of the intentions of actors. Using certain words and phrases calls into the mind of the listener a whole range of associated words and phrases. This may affect the audience's reaction, perhaps to the mystification of the speaker who is not touched by these associations.

Existing patterns of language and communication also impact on the operation of the other two variables identified: community attributes and material conditions. When Ostrom discusses community attributes, she refers to norms and a package that she describes as 'culture'. Clearly this must have a discursive dimension; the expression of culture uses discursive resources and, for some, culture and discourse are synonymous. Even aspects such as the distribution of resources cannot be treated as non-discursive. It may be a matter of visual checking to see if one person has more cows than another, but what are the discursive conventions by which one cow is counted more valuable than another? In many other cases, the value of resources has to be represented in some alphanumeric form; accounting, as Power (1997) has shown us, is a form of discursive practice. Finally, turning to material conditions brings us back to the social construction of reality and the need for perception and understanding of that reality to be expressed, particularly in situations of communication between actors.

One way in which analysts have sought to understand these processes of discursive patterning is through the concept of culture. Reviewing post-structuralist anthropology, Milton (1996: 62) distinguishes between two views of culture: 'people's activities and discernible patterns of action on the one hand, and what they are assumed to hold in their minds on the other'. She argues for a distinction to be maintained between these and for the concept of culture to relate only to the latter: 'It is also through culture that we reflect on our actions and experiences, describe them to others and plan future courses of action' (ibid. 63). Otherwise, she continues, it is not possible to explain how patterns of action are generated and to allow for 'the element of choice which is considered essential to the concept of action' (ibid.). This can be a difficult distinction to maintain, as perceptions are hardly separable from the actions they relate to, as Macnaghten and Urry (1998) emphasize. Thus others have seen culture as a way of describing the connection between representation or perception and social practices (Ingold 1993). This is closer to Habermas's concept of the lifeworld, described by Dryzek (2000: 22) as the place 'where meanings are negotiated and identities constructed by individuals'. Thus the level of patterns of language use, organized into discourses, provides a link between the culture of an organization or society and action.

Institutions and the Impact of Communication

E. Ostrom herself has paid some considerable attention to the ways in which institutions and communication are interdependent, notably in *Rules, Games and Common Pool Resources* with Gardner and Walker (1994). In this study, Ostrom *et al.* explore the role that communication can play in enabling co-operation in common-pool resource situations, doing so first through the medium of game theory and then through experimental evidence. This suggests a significant role for communication, particularly within environmental policy where co-operation over common-pool resources is concerned.

The key argument being addressed by Ostrom is whether facilitating communication is a worthwhile strategy in situations where common-pool resources are involved. Rational-choice theory, game modelling, and empirical experience all suggest that, in many cases, common-pool resources are depleted because of uncooperative behaviour and free-riding (see also Ch. 4). Ostrom, Gardner, and Walker (1994: 16–17) seek to explore whether communication can overcome these tendencies. They identify two distinct situations: those of learning and/or evolution where strategies change over time even though the basic incentive structures facing actors remain unchanged; and those where the rules-in-use are changed, leading to a new set of incentive structures. In the former case, communication can be an important aid to evolutionary change occurring; in the second case, communication is essential to establishing the new rules and sanctions, and also to operationalizing them, including promoting the understanding of these rules and sanctions.

On the basis of this study, they argue that communication can lead to agreements and commitments (that is, institutions) that prevent free-riding and resource depletion. They base this on laboratory experiments but argue that communication works because:

- it enables the offering and extraction of promises;
- it changes actors' expectations of each other;
- it can change the pay-off structure associated with resource use, particularly by threatening sanctions;

- it may reinforce prior normative orientations; and/or
- it can assist in the development of group identity.

They conclude that: 'The experiments provide strong evidence for the power of face-to-face communication in a repeated common pool resource dilemma where decisions are made privately' (ibid. 167). Interestingly, they also conclude that communication prevents excessive use of sanctions (ibid. 192–3).

These findings suggest the need for a close understanding of how communication occurs within environmental planning situations, which actors are involved, how communication is patterned, and what outcomes occur and are permissible. As will be seen, there is a range of literature available that has explored these aspects. However, Ostrom *et al.* fail to discuss exactly how the use of language in these situations of communication can make a difference. It is as if communication uses language transparently to achieve these changes. There are hints to the contrary as when the role of language in the form of verbal criticism or abuse is referred to as a ploy against defectors (ibid.). But the potentialities of this point are not taken up and explored further. Yet, all the transformative dimensions of communication listed above depend for their effect on the specific way in which language is used. This suggests a substantial lacuna in institutional analysis, which can only be filled by a closer attention to the use of language in communicative situations and a consideration of how the organization of arenas for communication affects such language use. This is, of course, closely linked to the individual discursive strategies of actors.

Institutions and Discursive Strategies

Language and communication has to be activated through the strategies of actors, through discursive strategies, that is, the intentional or purposive use of language within the policy process. There are a host of such strategies available, many of which will be referred to later in this book. In terms of the institutionalist framework, such strategies can be linked to the various ways in which the actor is specified. Four such dimensions were specified (see Fig. 3.1): resources; knowledge and information; valuations; and decision-making processes.

Resource use, as discussed by Ostrom, does not make reference to the use of discursive resources. This is a gap, since the arts of rhetoric and heresthetics (see Appendix and Riker 1986) show that language may be actively used by actors in a strategic manner; March and Olsen's recognition of the strategic nature of symbolic action is another example of this insight. Information and knowledge, as discussed in Ch. 2, are socially constructed and this must mean that there is some scope for the active construction of information and knowledge categories by actors. Similarly, valuations must be expressed and this poses choices as to how they are expressed. Finally, actors can seek to frame their decision-making processes (and those of others) in particular ways as an active strategy. Schön and Rein (1994), in particular, give prominence to the role of framing as an active strategy in promoting certain forms of professional decision-making (see Ch. 4).

Then there is the role of discourse in establishing and maintaining self-identity, including the perception of an actor's own interests. Here language is shaping and structuring actors, as opposed to being a resource involved in the actors' intentional use of language. As discussed above, the framing of valuations, information, and knowledge and of decision-making processes is all constructed through language and this can give rise to discursive strategies, but such construction is also involved in the reproduction of the actor as self-aware actor. This makes some links with Foucault's conception of the constructed nature of actors' subjectivities, but does so in a rather different framework, one in which the actor is conceived of as an entity beyond these subjectivities.

This is a complex process. Language is being identified as involved in all features of the IAD framework:

- actors' behaviour is discursively mediated or actually constitutes discursive strategies;
- actors' identity is discursively constituted;
- action situations are discursively mediated.

In all these ways, language is used and the prevailing patterns of language use have an impact. This is illustrated in the revision of E. Ostrom's framework in Fig. 3.3.

But while this emendment of the IAD framework identifies the ways in which language is involved, it does not go very far in telling us how it works. That would involve modelling a more strategic view of how language works and considering how discursive strategies operate. First, consider the options open to an actor in using language within the environmental policy process (or indeed any specific context). There is a vast range of linguistic resources potentially available to an actor. The set of what can physically be said is enormous; the set of what might be understood by another actor, although smaller, is still enormous. But a specific actor does not have this full range available to her. For any actor there is a subset of linguistic resources that is available: a given number of languages, dialects, and vocabularies. These may include, for example, English, French, economic jargon, middle-upper class dialects, and mathematical notation but not German, scientific jargon, Black English dialects, and Classical Greek references. Each individual actor has a very specific package of linguistic resources but, just as actors can be grouped into classes and other social categories, so the linguistic resources can be grouped into commonly held joint packages. This can aid analysis.

But of more significance to the analysis of environmental planning is the way in which such groups of linguistic resources influence policy processes and outcomes. To discuss this, one needs to consider how actors use language, and the assumption made here is that actors use language strategically. This means

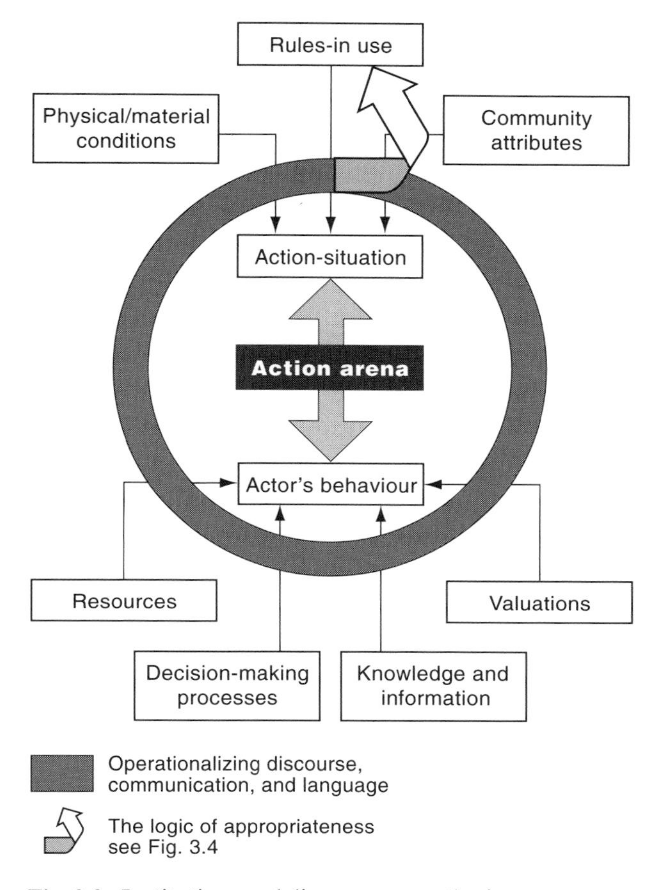

Fig. 3.3. Institutions and discourse: a synthesis

that actors use the set of linguistic resources that are available to them with a view to achieving a particular end. Just to clarify, this does not mean that the only ends to which actors are directed, when they use language, are material ones; it is perfectly possible, in line with Habermasian communicative reason, to use language with a view to achieving mutual understanding. However, this is still a case of the purposive use of language. It is in line with Habermas's view of language use in terms of argumentation, a view also adopted by Majone (1989) in his analysis of the policy process, and many other rhetoricians. Here language is seen as primarily involved with persuading another, whether in actual dialogue or implied dialogue, as with a speech or written text. Persuasion occurs through skilful argumentation so that language use is both end-directed and an accomplishment of the actor.

So language use is a purposive activity of the actor based on the set of linguistic resources that are available to that particular actor. This then raises the question of how the actor determines the appropriate linguistic strategy for the goals she has in mind. This involves a judgement about the context in which language is to be used. The aim of a linguistically skilful actor is to use the resources available to her in a manner most appropriate to the context, in order to achieve the given ends. How should this context be described? It is here that the concept of institutions returns. Institutions provide an account of the context within which language use occurs. And actors need to judge the requirements of these institutions when they choose their discursive strategies. The logic of appropriateness guides discursive behaviour, along with other aspects of behaviour. This is the case whether actors are seeking to advance their cause through compliance with an institution or whether they are actively seeking to disrupt or change that institution. In either case, actors need to take account of the institutional requirements when they use language to communicate.

Where actors judge their discursive strategies in terms of institutional requirements accurately, then they are more likely to achieve or at least move towards their goals. Of course, discursive strategies are not the only factor to take into account when analysing which particular outcomes arise. The IAD framework of E. Ostrom clearly identifies a number of other factors. But, in so far as the discursive strategy appears to have contributed to an actor's success, then this relates to its 'fitness for purpose'. And actors are likely to continue to use discursive strategies that turn out to be successful in terms of the actors' goals. There will be a reinforcing mechanism at work that encourages reliance on particular strategies. In this way, discourses—in terms of generalized patterns of discursive resources—are built up. These generalized patterns represent the heritage of strategies that actors have successfully used and feel they can rely on, and so are repeatedly used. Discourses then become part of the set of linguistic resources that actors draw on in developing specific discursive strategies. This new framework for considering discourse as discursive strategies, along with communication in forums and processes of discourse creation is summarized in Fig. 3.4.

Looked at in this way, it becomes clear why discourses generally fit with the institutional context and are likely to reinforce it. Individual strategies become discourses through repeated use and this only occurs where actors have found them effective. It is also clear why, in many cases, discourses support the position of powerful actors. Where actors are able to muster a collection of resources, both discursive and non-discursive, they are more likely to achieve their goals in terms of policy outcomes. The discursive strategy will be assumed to be part of that success and will enter the set of available resources to be drawn on again and again. Language, resource endowments, institutional context, and actors' judgements in fitting discursive (and other) strategies to the context are all part of the story by which actors achieve their ends, exercise power, and build discourses.

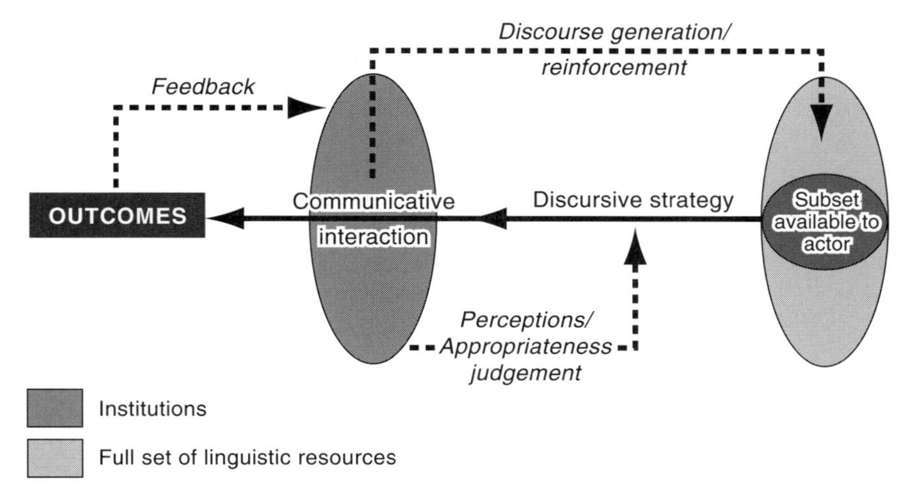

Fig. 3.4. Discursive strategies and institutions

Note: For simplicity this diagram has been drawn with reference to a single actor, clearly any situation of social interaction will involve multiple actors developing discursive strategies in this way and repeatedly going through the stages identified here, with subsequent feedback impacts.

Conclusion

This chapter has developed the argument that an institutional analysis provides the most appropriate basis for understanding policy practice and that this involves understanding the operational norms and rules-in-use of such practice. This provides an account, which recognizes structural patterns and also the role of the actor, actors' interests, and the social context. However, as formulated by the Ostroms among others, the institutional approach lacks a distinctive discursive dimension. Yet, within the institutional framework, all aspects of practice are clearly either linguistically mediated or linguistically constituted. Therefore, a framework can be developed that marries an institutional approach with this awareness of the importance of language use, identifying discursive patterns, looking at the interaction within communication, and considering actors' discursive strategies within such an institutional framework. This provides a fuller understanding of how discourse, communication, and language play a role within the policy process. This approach, and the selective use of the rhetorical method, will form the basis for the substantive analysis of environmental planning that now follows.

4

Conflict and Consensus in Environmental Planning

Having established a framework for analysing environmental planning in terms of institutions—a framework that allows for context and agency, the interests of actors, and a role for discourse—this chapter now moves into the substantive analysis of environmental planning. It begins by considering conflict and consensus-building in the environmental domain. There are a number of different approaches that have been taken to the existence of conflict over environmental issues. Such conflicts can be re-presented as issues of common interest. This is a common tactic within environmental planning and politics, as Ch. 1 demonstrated. Alternatively, an attempt can be made to bargain and negotiate on the basis of acknowledged conflicts, searching for a compromise solution. And finally, a more in-depth shift in the structure of interests can be sought, through techniques of collaboration and deliberation. This brings us back to the normative Habermasian theory, discussed in Ch. 2. Each of these approaches will be addressed, focusing on the discursive resources, communicative opportunities, and discursive strategies of actors as was highlighted in the institutionalist framework of Ch. 3.

The Prospects for Reframing Conflicts

Chapter 1 presented the influential discourse of common interests that shapes much of our general discussion of environmental issues. However, while a discourse of common interests may be a powerful resource at the general level, action in specific situations does not simply follow these broad discursive patterns. As the framework of Ch. 3 emphasizes, discourse is a resource to be used in these situations and will generally mediate action, but actors' behaviour and decisions, expressed discursively, are the motor that drives change. The discourse of common interests, therefore, has to be used as part of actors' discursive strategies. There is a difficulty here for common-interest discourses, since the framework of Ch. 3 made it clear that discursive strategies have to relate to the institutional context, including the norms of appropriate discourse but also the incentive structures facing actors. It is not, therefore, feasible for a discourse to be frequently adopted and activated during discursive strategies in direct opposition to prevailing incentives. This does not mean that all discursive strategies have to advance material self-interest, but they have to relate to an aspect of the set of incentives that helps define the actor's situation.

Actors, therefore, will search for discursive strategies and use discourses that relate in some way to their incentive structures, while at the same time seeking to adopt appropriate discursive behaviour. This is, of course, a recipe for using language insincerely. There are many social occasions when it is appropriate to use a particular linguistic formulation without actually meaning it. On a broader scale, there are many organizational occasions when a specific discourse has to be referred to as a matter of course, without it significantly affecting the other aspects of communication between actors. Such behaviour will have some effect because each discursive act is not isolated but connected to all other discursive acts through the common resources of discourses and language. There will be implications of making repeated use, even in passing, of a certain discourse with its associations, reference points, and rhetorical features. But these implications will lessen as the reference become less frequent, occupy a less central place within communicative interactions, and/or become more of a formality rather than a central act of social construction.

For these reasons it is quite possible for a discourse of common interests to become a reference point without defining or shaping the discursive acts of actors or achieving the task of reframing conflicts as common interests. In many circumstances such a reference has become almost ritualistic, a necessary feature of texts and addresses in order to count as appropriate behaviour but not a central defining aspect of the full text or address. It would take a change of circumstances, affecting actors' situations, for the discourse to be used differently. This must be highly frustrating for actors to whom the common-interest discourse has real resonance, who see the discourse being used formulaically but are unable to turn its rhetorical strengths into a broader currency for discursive acts. To understand this further, it is necessary to discuss the kind of discursive strategies that can be used to reframe environmental conflicts.

Reframing is a central concept in the work of Schön and Rein (1994). They use the concept to refer to the use of communicative strategies to achieve policy goals in situations of 'controversy'. Controversies arise because parties are operating with different 'frames', which then express themselves through different narratives and rhetoric. These frames are seen as grounded in institutions and operate at three levels. The most detailed level is the policy frame, which is used by institutional actors; this can take the form of a rhetorical frame—setting out the underlying persuasive use of story and argument in policy debate—or an action frame directly informing policy practices. Next comes the institutional frame and, at the most general level, there are metacultural frames, broad culturally shared systems of belief. Schön and Rein identify conflicts and lack of congruence between policy frames as the key factor underlying policy controversies and see the solution to such problems in 'reframing', a process which can be pursued by 'reflective practitioners'. Such practitioners draw on their store of lived work experience, not just in order to continue their daily practices but also to reflect on the successful and unsuccessful aspects of that experience in order to change future practice.

The central strategy that is adopted in reframing is the provision and exchange of information. This strategy is acknowledged within E. Ostrom's own framework, despite its general neglect of the discursive dimension. Indeed the rather strategic view of communication adopted by Ostrom and her co-workers tends to focus on reframing rather than any more extensive influence of discourse on the dynamics of environmental planning. E. Ostrom, Gardner, and Walker (1994) refer to the provision and exchange of information concerning rewards and sanctions accompanying particular courses of action and the future intentions of other actors. The idea is that such information provision will alter actors' perception of the action situation and prompt co-operative behaviour. This approach is close to the line taken by those within the risk communication school who see the provision of information—in the 'right' way—as altering people's attitude to risk and making them more willing to accept the location of facilities carrying environmental risks within their localities. Here, there is no actual change in the proposed course of events, but rather communication is oriented towards altering people's attitude to those events.

While the risk communication literature tends to focus on one particular type of environmental planning situation—the location of environmentally hazardous installations—its discussion of information provision provides a basis for assessing the discursive strategy of reframing. This approach places great emphasis on the information deficit facing the public and the need for education to fill the 'knowledge gap'. Traditional risk communication, therefore, limits its view of communication to this one-way traffic from expert to lay public, often referred to as the 'engineering' model of communication. Control of the process of communication remains with the knowledge-holders, the state or private corporate organization driving the process. Katz and Miller (1996) provide a case study of planning for low-level radioactive waste disposal in North Carolina in which there was considerable effort to provide information. This encompassed mailings, mobile resource centre, exhibits, meetings, free information phone line, news releases, newsletter, video-tapes, and so on. In her case study of the location of a waste incineration facility in Kimball, Nebraska, Belsten (1996: 40) highlights the way that the company, Waste Tech, provided full and open information, 'even to the point of providing interested community members with literature on the negative aspects of incineration technology.' The hope of such strategies is that full information disclosure and discussion will reassure and diffuse existing concerns rather than create more opposition.

However, there are clearly limits to this strategy. Information provision and 'remarketing' of new situations can only achieve so much. There are limits to the changes, the alteration to their lifestyles and life chances, that people will accept. There remains an institutional basis to conflicts of interest that cannot be endlessly reframed. This is a key weakness in Schön and Rein's original formulation. They argue that: 'Typically, the contestants in a symbolic contest enter into it on the basis of their interests in the policy process' (1994: 29) but go on to describe this relationship between interests and frames as 'reciprocal, but non-deterministic' (ibid.). In practice, they adopt a social-constructivist approach where the framing process is given priority over the pattern of interests and this enables them to see reframing as such a powerful tool and ignore its limitations. Perhaps this is because the cases they study are ones where economic interests are relatively peripheral; the case studies presented in their book all concern public policy issues and do not refer to the involvement of economic actors.

Unless there has been profound disinformation before (as in Hillier 1993 and Lauria and Soll 1996), the literature suggests that such information provision is likely to make a marginal change, but not fundamentally alter the nature of the situation. It may tilt the balance, but no more. And, indeed, if the process of providing information is mishandled and communities feel they are being manipulated, it may backfire and create doubt and distrust about the situation, amplifying the perception of costs and conflicts. In the Katz and Miller (1996: 123) case study, this is what happened: 'despite the best intentions of the Authority, the massive amount of material generated, and the enormous sums spent, this huge effort was judged by those to whom it was addressed as a failure'. This was because information and education went with 'seeming contempt for the public'.

There is a further problem with this strategy, relating to its treatment of different social groups and the way that information provision treats the preferences and valuations of these different social groups. The debates on environmental valuation have made it clear that the weighting of costs and benefits is not a context-free process. Critiques of such methods have pointed out how higher income groups are able to place a higher monetary value on a particular environmental asset they currently enjoy or an environmental externality they may be exposed to, because of the lower marginal utility of money for such groups. Without explicit counterweighting, any valuation exercise will intrinsically place more emphasis on the valuations of higher income groups (Jacobs 1997*a*).

But this is also the case in non-monetized situations. For the perception of costs and benefits is not the result of individual psychology, a preference structure to be found somewhere in people's brains or psyches. Costs and benefits are socially constructed by actors within specific circumstances (Macnaghten and Urry 1998). Thus the costs of locating a waste treatment facility within a lower-income community will be perceived differently by that community than would be the case with a higher-income community. To make it quite clear, this is not because lower-income groups value their health or environment less than higher-income groups. They may well care just as much or more about the environment, their children, and their neighbours. Rather the circumstances mean that these valuations only function in the context of other valuations, concerns, and worries, and for lower-income groups these will probably be more basic material matters of employment, housing, education, and existing health. The consequence is that strategies for presenting information and alternative perspectives on costs and benefits will impact differently in higher- and lower-income communities.

For example, Weaver (1996) tells of how the campaign to protect the Smoky Mountains, USA, from development was unable to cope with the self-perceived reality of the people who live in the area. The campaign was framed to build a consensus among high-income groups concerned with the area but had no resonance with low-income mountain residents. Furthermore, the rhetoric of the campaign built on negative stereotypes of mountain peoples and developed from initial condescension to rhetorical abuse. A similar analysis is provided by Senecah (1996) of the campaign to protect the Adirondacks, USA. Here local low-income groups were made to feel invisible by the environmental arguments, which focused on 'natural' features: 'the environmental groups frame their appeals around those very aspects of preserving open space that reinforce the siege mentality of the opposition groups and reinforce the invisibility of the Adirondackers' (ibid. 111). In the event, this led to conflict over environmental regulations and an alliance between private property interests (who were opposing protection to ensure their development rights) and the lower-income residents; the local people had been empowered by the 'wrong' side.

Of course, theorists of deliberative democracy or collaborative planning (discussed in Ch. 2) would deny that genuine interaction is occurring in these exercises. They would argue that the limitations to this strategy of reframing lie in the one-way nature of the communication. So often, as in most of the risk communication case studies, it is a matter of trying to persuade the community who will bear the costs, that those costs are not as serious as they fear. Even where the communication runs two ways, so that the instigators of the change (the developers, say) begin to reassess their own costs and benefits because of new information provided by the community, this does not represent an interactive process. The discussion now turns to strategies of negotiation, collaboration, and deliberation to shed more general light on how conflicts can be discursively handled in a more interactive manner. Reframing still has a part to play but it cannot be a sufficient account.

Discursive Strategies for Negotiation and Bargaining

Putnam and Roloff (1992: 2–3) distinguish negotiation and bargaining from either persuasion (of the type involved in reframing) or joint decision-making (as in collaboration and deliberation). They describe it thus:

Bargaining entails two or more interdependent parties who perceive incompatible goals and engage in social interaction to reach a mutually satisfactory outcome . . . Each party in the relationship must cooperate to reach his or her objective and each party can block the other one from attaining his or her goal . . . This interdependence, combined with potentially antithetical goals and demands, sets up a mixed-motive relationship in which both parties cooperate by competing for divergent ends. Rules and normative practices of bargaining include specifying preferred outcomes prior to the negotiation, exchanging proposals and counterproposals, and engaging in dynamic movement through social interaction . . .

In negotiation, the communicative interaction between actors results in actual changes to the outcomes of the situation so that benefits are added and/or costs reduced. This is not just a matter of revising the proposal or plan on paper. As anyone who has engaged in any sort of negotiation knows, negotiation is a highly complex communicative act, or series of acts. It involves explanation not just about the desired outcomes that each actor wishes to achieve, but also evocation of the relative value of different outcomes. For negotiation is not about each actor achieving his or her own desired outcome. It is about achieving a compromise, an acceptable package of outcomes, which will probably fall short of either actor's ideal. Nevertheless, actors involved in negotiation do so on the basis of their perceived self-interest and negotiate because they consider their interests are best served in this way (Bacow and Wheeler 1994: p. vii). This in itself requires an understanding of each actor's standing within the negotiations and their right to speak for their own preferences and desires and to expect some degree of accommodation of those preferences and desires. The discursive construction of *locus standi* for each participant to the negotiation is a vital first step. Hillier (1998) helpfully distinguishes between antagonism—adversarial conflict entailing mistrust and suspicion—and agonism—the constructive mobilization of differences towards decisions that are partly con-sensual (*sic*) but that also accept unresolvable disagreements.

There is a well-established 'how-to' literature on negotiation, available to practitioners. It puts forward proposals for resolving disputes through a set of tried-and-tested procedures for managing negotiation, including techniques such as mediation. As such, it is not limited to the field of environmental policy; for example, mediation is widely used within family and legal disputes. But conflicts over the location of land use with adverse, or at least debatable, environmental impacts have also proved a fruitful area for applying such skills (Crowfoot and Wondolleck 1990; Bacow and Wheeler 1994; Glasbergen 1995). The keyword here is 'skills'. This literature draws on the past experience of and is addressed to practitioners. It emphasizes the details of their practice and makes recommendations for improving it. It is focused around the idea that practitioners can acquire new skills and, thereby, affect the policy process. In doing so, it emphasizes the detail of how practitioners exercise their skills—the forums they choose for bringing interested actors or stakeholders together, the procedures chosen for governing the interaction of these actors and the ways in which that interaction occurs. Above all, there is an emphasis on how actors talk to each other and how this can be managed.

The difficulty with this literature is that it is essentially a heuristic methodology for developing policy guidelines. It doesn't offer (and does not pretend to offer) any generalized theory of the policy process or the role of discourse within that process. Instead it relies heavily on unsituated examples of past practice. This makes it difficult to predict when the recommendations for revised practice will work and when they will not. There is no account of the broader constraints that will affect policy practice in any particular case. Instead this is generally an optimistic literature that suggests that it is worthwhile trying to adopt revised procedures and practices. Actors' interests are obviously seen as significant for structuring the negotiation process but they are not seen as inevitable constraints on outcomes. The emphasis is rather on seeing how negotiation activities and the skill of the practitioner can identify win–win scenarios and help parties to see their interests in a new light, offering new compromises. Furthermore, such approaches see language as a highly malleable system. Language is subordinate to decisions by actors about the nature of their communication; it is a transparent medium and the emphasis falls instead on the organization of forums and the sequencing of interactions.

However, explicitly considering the discursive strategies that can be adopted to move the negotiations forward reveals a range of tactics involving language. These are as diverse as emotive pleas to concede a particular point and poker-faced diversionary tactics, which allow a desired outcome to slip through the net of negotiations unnoticed. The influence of emotive pleas can be difficult to trace as often public policy officials are unwilling to admit to being swayed by such 'non-rational' concerns. But they do play a discursive role. By tracing the chronology of discussion in the case of an International Joint Commission's hearing on the water quality of the Great Lakes, Waddell (1996: 148) shows how the Commission were influenced by emotional public testimony relating to cancer fears, although they subsequently claimed that this was only because it was corroborated by scientific findings.

Aside from such emotive pleas, bargaining also centrally involves the use of offers and threats to achieve concessions and compromises. Tutzauer (1992) has examined the linguistic features of offers. He specifies them as usually numerical in nature (or capable of being translated into a scale); demanding of a response (and hence linguistically paired with that response, even a nil response); and fluid (that is tentative in nature and capable of change in expression). By contrast, threats have been characterized along five dimensions (Gibbons, Bradac and Busch 1992).

First, there is the common use of polarized language, which sets one party in opposition to another. This can be modified by other features, such as the level of verbal immediacy, which conveys the speaker's feelings through specifying the intended distance from the other party, with low immediacy implying greater distance and dislike. Then there is the intensity of language used, which can signal the positive or negative direction of the negotiations. As Gibbons *et al.* (ibid. 164) point out, 'Language intensity seems particularly salient for conveying the credibility of a threat or promise and for assessing the intentionality of threats in negotiations.' In an American context, they identify the use of profanity, sex, and death metaphors with high intensity. The fourth dimension is lexical diversity, the range and richness of vocabulary; high diversity is associated with confidence and competence in negotiations, low diversity with anxiety. However, Gibbons *et al.* also note that repetition brings with it low lexical diversity and this can be an indicator of firm commitment and toughness. Bringing levels of diversity into parallel with the other party can also be a sign of approaching agreement. And finally, there is the dimension of language style and, in particular, the use of a high-power style; again a high-power style conveys certainty and commitment, attributes of a strong bargaining position, although a low-power style may be used tactically to convey lack of room to manœuvre.

This detailed research on negotiation illustrates how linguistic features can make a difference to the bargaining process. Such research has tended to concentrate on the adversarial nature of that process; this leaves a central element of negotiative processes underdeveloped. This element is trust. Within negotiation, the creation and nurturing of trust is essential because many of the revised outcomes that arise from negotiation will only be delivered in the future. It requires trust by all parties that all will stand by the agreement reached. Using an argumentation approach, Keough (1992) found that negotiation involves two distinct but interrelated activities: instrumental functions, which focus on achieving aims; and relational/identity management functions, which seek to build trust, promote unity, and manage power relations. A central element of the relational function is the provision of prenegotiation accounts (ibid. 118): 'An *account* is an explanation that justifies a decision (i.e. reasons given in support of a claim). Its purpose is to manage conflict before it happens . . . Its effectiveness depends on the timing of delivery, the perceived adequacy of the account, and the perceived sincerity of the account giver.' Such accounts encompass the causal (explanations that try to mitigate responsibility for unfavourable outcomes), referential (attempts to reframe the consequences of undesirable outcomes), and ideological (appeals to shared values, belief systems, or joint goals). In relation to the last Keough notes that 'the search for shared values is laudable but not always plausible' (ibid.).

This means that negotiating processes are characterized by two rather different discourses: the adversarial bargaining of threats, offers, bluffs, and pleas on the one hand; and the language of trust creation, building confidence in relationships between parties, on the other. The roller-coaster movement between these two discourses characterizes negotiation, the aim being to end with a mutually agreed compromise in an atmosphere (and couched in a language) of trust. However, during negotiation, trust is an instrumental resource, enabling the bargaining to proceed and conclude satisfactorily. It is not the ultimate goal in itself, and it can often be a strictly constrained form of trust that operates, supported by legal documentation and the presence of a regulator. The emphasis tends to lie with the more adversarial language of bargaining. There are a number of reasons why a discourse of conflict tends to dominate in negotiation, despite the strategic need for trust at certain points during the process.

First, not all negotiations are conducted face to face. Poole, Shannon, and DeSanctis (1992), reviewing the literature on the role of communication media used within negotiation, found that the media affects issue definition, the search for a solution, self-presentation of parties, the climate of negotiation, and the management of the balance of power. Comparing face-to-face encounters with the use of text, audio, and video media to convey communications, they found that face-to-face contact is less likely to bring conflicts to the surface, create we–they oppositions, and allow negative expression of emotions. It is more likely to avoid escalation of conflict and emphasize difference of status between parties (ibid. 60–1).

Second, it is to the advantage of each party to negotiation to appear in a strong position. Where negotiation occurs in the context of regulation and the threat of a regulatory veto, then those opposing the situation have an incentive to appear fairly intransigent, to emphasize the zero-sum nature of the existing situation. By doing so, opponents strengthen their hand in negotiations and make any concessions achieved appear more valuable. But the same tactic, of course, will work for the proposer or developer; if they appear intransigent, the concessions that they offer will also appear more significant. So the maintenance of an appearance of conflict is an appropriate discursive strategy for parties to a negotiation, even though this can undermine the necessary trust for successfully concluding the bargain.

Third, there is the need to maintain the internal solidarity of each actor. So far, actors have been referred to as if they are individuals, even if collective individuals—the community, the developer. However, in practice, these actors are themselves collectivities that need to maintain solidarity. Within organizations, whether public- or private-sector, this is maintained by managerial structures and the ultimate threat of dismissal or relocation for any individual who 'breaks ranks'. Within the voluntary sector, though, other mechanisms need to be brought into play to maintain solidarity. For voluntary collective action always threatens to split apart under the pressure of the costs of joining in and the benefits of free-riding on someone else's action. This is one of the key insights of institutional rational choice as applied to community activism. For most members of a community, the costs of engaging in collective action outweigh the anticipated benefits (Olson 1969). Costs are certain and happen here and now; benefits are potential and occur in the future. The involvement of each individual is unlikely to make a significant marginal improvement in the chances of achieving these already uncertain future benefits. So each individual feels reluctant to join in and bear the resultant costs. Following this line of reasoning, collective action within a community is always likely to be constrained.

Furthermore, collective action is more likely to occur within communities which can reduce the costs of joining in and/or increase the chances of achieving satisfactory outcomes. Therefore, activism is more likely to be found among the retired or unemployed, with a low opportunity cost for their time, and among those with the resources, contacts, knowledge, and skills to increase the probability of their activism having an impact. This describes closely the composition of many middle-class dominated action groups and can help explain the lack of mobilization among poorer groups (Hillier and Van Looij 1997). There are other factors to consider, of course. Strong values will also result in a discounting of time spent in activism and some 'joiners' will count the social contact with others a benefit, either in terms of social activity or as a means of reinforcing social identity. But these factors only modify the underlying collective action problem; they do not replace it as an explanation.

Having made such an analysis, it is possible to derive strategies for encouraging collective action within the voluntary sector (Rydin and Sommer 1999; Rydin and Pennington 2000; Rydin 2000). These include: reducing the costs of participation in collective action (e.g. paying for childcare, keeping events local and short); increasing the direct benefits by making events more enjoyable; explaining, increasing, and highlighting the impact of action on policy outcomes; focusing on issues where benefits are local, visible, and immediate; and penalizing non-participation and free-riding (e.g. through 'naming and shaming'). It is notable that many of these strategies have a discursive dimension, particularly those concerned with identifying the linkage from collective action to policy, marketing the benefits of collective action, and penalizing non-participation.

But it is also notable that one of the easiest ways to enhance solidarity amongst those who do undertake action is to highlight the conflict of the zero-sum situation. The discourse of conflict becomes a way of achieving group coherence, rather than fostering dialogue with opponents. Ingham's (1996) study of planning for the small community of Red Lodge, Montana, shows how dissensus, rather than the pursuit of consensus, played an important role in galvanizing the community, bringing them together, and encouraging them to learn more about the development issues in the locality. She points out that the reason was possibly that the community was only just beginning to grapple with the issues, and that consensus-building might be more appropriate at a later date, but she argues that at this stage the conflict was actually beneficial. Cantrill (1996) makes a similar point in relation to the Beartooth Alliance, which opposed mining at Cooke City, within Yellowstone National Park. Here a rhetoric of location, distrust, history, and hardships helped to foster collective action by the community, but only at the cost of cementing their self-identity as a marginalized group. Cantrill concludes that this prevented collaborative planning in the public interest, but the point to note here is the way that a rhetoric of conflict reinforced self-identity and collective action by communities. In his study of 'Image Politics', De Luca (1999) also points out how rhetoric can help constitute the identity of protest groups and aid in their mobilization. He sees the rhetoric and the collective action itself as constitutive of the identity of these groups (see also Bridger 1996).

The basic problem here is that the factors that bind a group together and create more favourable conditions for collective action are also those that distinguish that group from others. For example, in the case of the proposed coastal super-quarries in Scotland and Nova Scotia, Dalby and Mackenzie (1997) highlight the way that community identity and cohesion was reinforced by specifying the external threat and thereby specifying what was threatened (see also Wilson 1996). This makes even simple negotiation difficult, let alone more ambitious aims for collaboration, joint decision-making, and deliberative democracy (see below). Burningham (2000) highlights one aspect of this in her study of the use of the language of NIMBYism. She shows how parties to a development dispute are very aware of the consequences of being labelled 'NIMBY' and seek to avoid it; however, part of their tactics for doing so is to characterize their opponents as 'NIMBY'. Her main point is that claims based on NIMBYism are seen as less legitimate than those based on wider environmental concerns. This is a debatable point but her work does show how conflict-reinforcing claims are actively used by parties to a dispute in a way that will undermine the potential for negotiation.

One response to this tendency for conflict to remain discursively entrenched in negotiation situations is to use mediation to recreate an atmosphere of trust (Troja 2001). This seeks to use an independent third party to help the parties come to an agreement. The mediator works by directing the bargaining and discussion, and by simply keeping the process going and preventing it from stalling. The role of mediators is central as they have a vested interest in keeping the process going until it is successful. Mediation provides an externally generated structure, and as a consequence parties to negotiation are less likely to feel that the negotiation situation is biased against them. Indeed a mediator or facilitator can help avoid systematic bias arising during negotiations (Susskind and McMahon 1985). And mediation, as a whole, is strongly goal-oriented, the goal being the achievement of an agreed compromise. However, the appointment of a mediator cannot fundamentally alter the dynamics associated with actors' incentive structures. Holzinger's (2001) study of failed alternative dispute-resolution techniques in Germany shows how external parameters can limit the scope for negotiation. What mediation can do is extend and multiply the moments when trust is generated and when it can be used instrumentally to support the negotiation of an agreement. Some forms of mediation can so multiply these moments as to approximate to deliberation and collaboration. Such deliberation/collaboration is seen as a response to the persistence of conflicts and the inability to produce mutually acceptable outcomes. They are discussed next.

Collaboration and Deliberation

If negotiation tends to be dominated by the discourse of conflict, despite the need to generate trust (at crucial times), how does this differ from processes of collaboration and deliberation? These processes are not just extensions of negotiation and bargaining. Deliberation and collaboration represent a breakpoint with negotiation. As deliberative democracy theorists claim, such collaboration—if successful—would be a distinctive form of governance and not just a problem-solving technique for a particular conflict situation. Collaboration is not, therefore, just a way of turning a specific zero-sum situation into a positive-sum one, but a way of creating the social conditions whereby positive-sum situations become the norm. Within collaboration, the whole situation becomes open for reconsideration so that the outcomes can be altered much more substantially than with compensation, information-provision, or bargaining. New outcomes are posited, based on new knowledge, joint thinking, and previously unconsidered possibilities. In effect, the situation becomes jointly redesigned with a view to providing benefits for all parties to the collaboration.

There are various examples of contemporary planning practice influenced by ideas of collaboration and deliberation. For example, citizens' juries are becoming more widely used (Crosby 1995; Smith and Wales 2000). These typically involve a group of between twelve and twenty-five people, chosen to represent various positions within the affected community, who will meet together over a period of two to four days. During that time they will be presented with information, some of it via expert witnesses. There will be the opportunity to question and argue with these witnesses and—most importantly—time to deliberate together over the final decision, recommendation, or other outcome. The idea is that members of the citizens' jury have the opportunity to think about their views and preferences and, indeed, that these will be formed or changed during the deliberative process. All jurors are expected to play a part and no one opinion is allowed to dominate the proceedings. The outcome should not be a compromise nor, in some way, a mediation between the different views expressed, but rather should encapsulate a form of rationality, leading to better decision-making. Other examples include visioning exercises, 'Planning for Real' role-playing, consensus-building, and consensus conferences.

One example of the influence of ideas of deliberation and collaboration is to be found in the contemporary risk-communication literature. As compared to the old traditional 'engineering' model, more recent thinking in risk communication now has at its core a normative commitment to greater public involvement in and even control over environmental risk management. Belsten (1996) provides a case study of exemplary risk communication practice in which 'open, inclusive, broad-based, ongoing rational discourse' enabled the successful planning of a hazardous waste incinerator in Kimball, Nebraska. There was an emphasis on broad participation, not just the involvement of experts; on the experiences of 'real people', not just statistics; on concepts such as 'rights', 'morality', 'social responsibility', 'justice', and 'obligations', and not just 'cost-effectiveness' and 'efficiency'. But, as she makes clear, the key factor was that the community had the option to say 'no' to the incinerator; they retained control over the decision. While this example highlights the potential of the new paradigm of communicative risk management, it also points to a limitation. As with the literature on mediation, great faith is placed on the ability of communication to overcome conflict. The test of the new risk communication is whether it works not just when both sides *agree* on the location of an incinerator, but when they continue to *disagree*.

A commitment to deliberation clearly requires a completely different language from negotiation. As Bohman (1996: 41) says: 'deliberation requires a special form of communication that begins when extant forms of communication and shared understandings are strained or even break down'. The discourse of trust must become the dominant discourse, replacing that of conflict. But, as has been shown, this involves a degree of sacrifice for actors, since this discourse of conflict was supportive of internal solidarity and a strong external bargaining position. Making this shift can, therefore, be expected to be very difficult. However, it is not possible to substantiate this in practice as there are very few real-life examples of deliberation and collaboration to study. The literature is, as Smith and Wales (2000) point out, largely theoretical in nature. This remains mainly a proposed way of dealing with conflicts. The discussion of discourse within deliberation, therefore, has to draw on the normative literature for suggestions as to how the language of policy-making would need to change. This involves abstracting from the, at times, utopian ideas about the conditions required for changed policy practice. But such abstraction will be combined in the discussion here with examples of collaboration and deliberation in practice.

First, it is necessary to clarify the nature of deliberation and collaboration. The normative literature has increasingly come to focus on deliberation (rather than the looser collaboration) as a form of joint decision-making in which the reasons for the decisions are, in some sense, more rational and the outcomes more fair. Bohman (1996) provides three different models of such deliberation. A precommitment model assumes a prior agreement or consensus on the values of deliberation; a procedural model focuses on the procedures, forums, and processes that can be designed to promote deliberation; and a dialogic model sees deliberation as a process of giving and, more importantly, exchanging reasons iteratively. Bohman argues for a dialogic model based on criticisms of the precommitment model as being utopian (which is surely correct) and of the procedural model as tending towards a reliance on constitutional change within the state (a point that Dryzek 2000 agrees with).

The reliance on reason-giving as the central defining feature within deliberation need not mean that deliberation has to be based on rational, logical forms of reasoning. Critics have pointed out that such reasoning is likely to be highly divisive as it privileges groups with particular reasoning skills. Rather both Bohman and Dryzek (along with others) have emphasized that reason-giving within deliberation can take a variety of forms: argument, rhetoric, humour, emotion, testimony or storytelling, or gossip; that is, any non-coercive mode of communication. Green (1999) puts particular emphasis on the role of the cultural arts. Dryzek would, however, apply two tests for the admission of any communication: that it is non-coercive and is able to connect the particular to the general, thus avoiding purely self-referential autobiography.

Bohman echoes Dryzek's concerns but, rather than apply normative tests, he identifies five dialogic mechanisms that give a hint of the type of language that should dominate if deliberation is to work effectively. The first is that speakers need 'to make explicit what is latent in their common understandings, shared intuitions and ongoing activities' (1996: 59). Bohman suggests that this works best by using an already accepted interpretative framework, and modifying it. But he does point out that modifying such frameworks and conducting such communication in public also has implications. One is that the authority of individual speakers is reduced, as compared to their situation within established contexts; the other is that the constant demands for interpretation that result will undermine culturally specific vocabularies and disperse specialized vocabularies more widely (ibid. 43–4). Healey (1996: 225, 1997: 274) also makes the point that, before this occurs, allowance has to be made for translation of multiple languages within collaborative situations.

In an example of a quasi-deliberative process around waste management in Berlin in 1997–8, Troja (2001) examines a series of twelve forum meetings and two working groups that sought to involve local stakeholders in developing an improved strategy. He emphasizes how these stakeholders differed in their way of reasoning and style of argumentation, and the demands this placed on the process. Such an approach involved evaluating very different types of knowledge that the various stakeholders brought to the process, so that much more than mere translation was involved here. In this particular case the outcome was not a consensus but a greater willingness among parties to accept and live with the expressed differences of view. And van Woerkum (2002) has argued that more than translation is involved in the sense that groups may work with quite different types of orality or oral culture (see Chs. 6 and 7 for further discussion of this work).

Second, rather than rely on shared values and commitment, speakers should engage in 'back-and-forth exchanges around differences in biographical and collective historical experiences' (Bohman 1996: 60). This echoes Green's emphasis on memories and hopes in building communication between communities. Such exchanges are important because they can help build solidarity and also mutual recognition. At root they are based on recognition of the equality of each actor's position, situation, and perspective, recognition linked to mutual respect. Healey (1996: 226) also emphasizes the need for respect. Such mutual respect has to be reflected in the language of deliberation: the shift from 'I' and 'you' to 'we' (Green 1999: 44). Green (ibid. 210–12) draws on her experience in developing a community vision in a neighbourhood of Seattle, USA. Here a successful visioning experience was based, not only on shared expectations and anxieties and a community of limited heterogeneity, but also on mutual education 'to establish common ground about the history of their shared place'. In this case study, Green notes both the 'predictable tensions, struggles for dominance, rival priorities' *and* the respect with which speakers treated each other, the general ethos of calm and the high level of trust. She contrasts this with the King County Organizing Project, a broader coalition aimed at community development, where the language of 'selling' dominated within an ethos of justice-oriented urgency. This failure to 'walk the talk' of deliberative democracy was considered responsible for the failure to maintain a high level of trust in this case (ibid. 214).

An interesting point here is where the new boundaries of the 'we' are drawn. For while 'we' is more inclusive than 'I', it is still defined with reference to 'you' or 'they'. In a study of the Swan Valley, Perth, Healey and Hillier (1996) provide an analysis of a conflict between local communities and the state planners, who were willing to consult but not negotiate, let alone collaborate. In an attempt to force the planners into a more collaborative style, the community engaged in collective action. This involved reinforcing the identity of the community and this, in turn, involved building bridges between the local white working-class and aboriginal communities. In order to do this a shared construction of their identity and history developed. 'Them' and 'us' became 'we'. But it was essentially a strategic action, although it may have had richer side effects. One wonders what would happen to the communities and their self-construction once the stance of the planners changed and some form of negotiation commenced.

The third mechanism particularly applies to policy issues where a general principle is being applied in a specific case; here speakers need to be able to move back and forth between 'a general norm and its concrete specification' (ibid. 61). Fourth, and related to this third dialogic mechanism, speakers also need to be able to move back and forth between 'a vague and abstract ideal and various proposals to make it richer and more comprehensive' (ibid. 62). Together, these two mechanisms emphasize the dialectical nature of deliberation, the give-and-take, the back-and-forth.

Finally, there is often within deliberation scope for 'perspective taking and role taking' (ibid. 63). Green (1999: 23–5) gives this particularly emphasis in her proposals for deep democracy. She highlights the use by Habermas of Mead's discussion of role taking. Again, the adoption of different roles, particularly the taking of another's role, is seen as a way of breaking down barriers between parties to deliberation and fostering co-operative action. Green hopes for the possibility of 'an "ideally extended we-perspective" that authentically represents an "interlocking of perspectives"' and that this '*could* foster a broader sense of community amidst valued differences within a desirably pluralized global lifeworld' (ibid. 24–5). In a study of various planning processes in the California area concerning water, transport, and urban development, Innes and Booher (1996) have shown how role-taking is highly significant in moving the discussion forward. Individual parties to the deliberation take on a variety of roles, moving between them without effort. They share and exchange roles, as well as playing out their own repertoire. The dramatic nature of such communication is highlighted and this requires a degree of acceptance: acceptance of mobility of roles, of inconsistency in an actor's position over time, and of variability in the language used by any one actor.

This discussion provides an outline of how the discourse of deliberation might look. The key discursive features of deliberation appear to be an expression of divergent perspectives and the perception of conflict, but contained within a discursive framework of mutual trust and respect. While some commentators see deliberation in terms of a stable atmosphere of calm that is potentially threatened by the expression of conflict (Smith and Wales 2000), in fact deliberation has to provide the opportunity for such expression of conflict in a variety of 'safe' ways, including role-play. Successful deliberation has to overcome the tendency for face-to-face encounters to suppress conflicts (see Poole, Shannon, and DeSanctis 1992), rather than accept this as a quick route towards a compromise. Similarly Edelman (1988: 25) points out that deliberation requires a much clearer recognition of difference than in negotiation, where ambiguous language about interests and positions can actually facilitate bargaining.

Deliberation involves experiment. It is both unpredictable in its progress and outcomes, and unsettling for participants. Green (1999: 42) describes it as 'both *deeply disruptive* and *deeply beneficial*'. In surveying the general use of arguments within joint decision-making, Keough (1992: 116–17) finds evidence to support the role of apparently critical and disruptive behaviour in advancing collaboration. She finds that effective joint decision-making is based on assessing each other's arguments, rather than just accepting them. The key to success here is not consensus but building a position based on divergent perspectives. This contradicts those who hope that deliberation will lead to convergence of underlying values and preferences (e.g. Jacobs 1997*a*: 221). It does, however, emphasize that deliberative discourse has to be oriented to explaining these divergent perspectives and not just defending them. Innes and Booher (1996) found humour to be an essential element of handling the highly charged and often emotional situations that arise.

This discussion raises some key questions for proponents of collaboration and deliberation. Much of the literature has focused, explicitly or implicitly, on the interaction between different communities or groupings within civil society. It remains untested how interactions with economic actors can be integrated into this model. Some, such as Healey, remain optimistic that a variety of actors can be brought into the deliberation process. But can economic actors engage in the variety of role-play and flexible give and take of dialogic deliberation? This remains a contested point (Tewdwr-Jones and Allmendinger 1998). Certainly, proponents of Local Agenda 21 with its more flexible forms of debate and discussion have reported an unwillingness of economic actors to engage with such new modes of communication (Darlow and Newby 1997; Miranda and Hordijk 1998; see also Ch. 9 for further discussion of LA21).

A similar point arises with state actors. Deliberation of the type outlined above would be highly disruptive of existing rationalities of the policy process. Can policy-makers handle the risk that would come with deliberation? These issues are explored further in Chs. 5 and 6, but at this stage it is appropriate to raise the question of whether the state and policy officials have any interest in promoting deliberation. One possibility is that the state would see deliberation as a way out of specific crises of legitimation. This would certainly explain why deliberative-type modes have been adopted in environmental policy, where particularly intractable problems such as the disposal of waste have been used as test-cases for deliberative exercises. It would also, though, suggest that these are unlikely to become either a recurrent feature of policy practice or to have much influence on actual policy outcomes and decisions.

Dryzek (2000: 29) fears that deliberative processes could become co-opted into the rationality of the state. Rather pessimistically he points out that the 'state is increasingly subject to the constraints imposed by the transnational capitalist political economy' and that 'public officials under the sway of such imperatives are highly constrained when it comes to the terms of the arguments they can accept; [that] it is very hard for deliberation to reach them'. He therefore has come to see deliberation as remaining mainly a feature of civil society or the public sphere. Dryzek (ibid. 50–5) suggests that the contestation of discourses and their communication to the state via rhetoric can be used to influence the state from outside. However, his model of contested discourses remains very general. Rather surprisingly in this formulation, successful deliberative democracy is seen as little more than the reframing of issues, with environmental risks being reconstructed as justice issues (ibid. 77) and whaling being affected by a changing construction of 'what is a whale?' (ibid. 126).

This model is rather close to a form of discursive pluralism. Dryzek sees the democracy of contestation arising from a rather unspecified network form of control and management. This ignores the many constraints that operate on the circulation of discourses. It also ignores how the interaction of discourses is shaped by their discursive structure and how this influences the outcomes of discursive contestation. Finally, Dryzek's suggestion leaves groups in the position of engaging in strategies of lobbying and resistance, rather than fulfilling the hopes of deliberative democracy as a new mode of governance. His justification is that 'there have to be moments of decisive collective action, and in contemporary societies it is mainly (but not only) the state that has this capacity' (ibid. 79).

This throws deliberative democracy back into the arena of political activism. As such, it becomes vulnerable to the collective action problem, discussed above in the context of negotiation. As Hillier and Van Looij (1997: 21) state, 'inclusionary argumentation involves more than giving people "accomodative voice"'. Green (1999: 80) hopes that, once tried out, the parties to deliberation will see so many benefits flowing that they will readily sponsor its spread. But how does the first successful deliberation in a particular policy context take place? How is the prisoners' dilemma overcome, which would otherwise frustrate people's involvement in deliberation on the basis that no individual could see sufficient benefits arising to outweigh their own costs of joining in? Again the unsettling and disruptive aspects of deliberation have to be taken into account; they are likely to impact well before the benefits of deliberation are seen. If deliberation can be seen to deliver benefits for actors, then that may be a real incentive to become involved, but without demonstrable benefits, these have to be taken on trust.

However, there is some scope for more hope. It is possible that repeated interactions, the generation of trust and reciprocity can actually alter incentive structures facing actors in a more profound way and, therefore, that discursive strategies focused on these features will over time support the long-term embedding of collaborative arrangements. The suggestion that repeated interactions, trust, and reciprocity can alter reactions to incentive structures, even change the incentive structures themselves, draws on the growing literature on social capital (see Rydin and Pennington 2000 for a review in the context of public participation strategies). Ostrom's (1990) work, in particular, has shown how local communities can develop institutional arrangements for acting co-operatively. Here there is an emphasis on the ways in which information, relationships, and trust are articulated through social networks. Reputation and its relationship to the development of norms of reciprocity are particular concerns. The desire of an individual or group to maintain their reputation in a close-knit social context may lead to the development of co-operative and collaborative behaviour; as Chong (1991) argues, reputations can be considered as general commitment devices, rationally conceived, but relatively unconsciously or habitually followed. Repeated interaction and communication in small-group settings also allows individuals and groups to monitor one another and to exact sanctions on non-cooperative behaviour. In these circumstances, it is possible for people to develop a reputation, to learn whom to trust, to learn what effects their actions will have on the achievement of common goals, and to discover how to work collaboratively in order to gain benefits and avoid harms.

What is involved here is the creation of 'social capital'. Used in its broadest sense, the term 'capital' refers to those goods or ideas with which something else can be created or established. Social capital, therefore, constitutes the pre-existing elements of social structures, which social actors can use to obtain their objectives. As Coleman (1988: 98) points out, the existence of social capital facilitates 'the achievement of certain ends that in its absence would not be possible'. More concretely, social capital encompasses such things as:

- the extent of networks between individuals and groups
- the density of relationships within networks
- knowledge of relationships within networks
- the existence of obligations and expectations regarding these relationships; i.e. promoting reciprocity
- other forms of local knowledge
- the level of trust between individuals and groups
- norms of routine behaviour
- the existence and use of effective sanctions to punish free-riding.

These dimensions of social capital can be translated into the costs and benefits of individual actor's incentive structures, but such a calculation misses the way in which these various institutional arrangements work. It is true that the penalty for not engaging in local co-operative activity will be weighed against the other benefits of non-cooperation by an actor; but additional money, additional prestige, and feelings of shame do not all impinge on an actor in the same way. Depending on the local cultural dynamics, shame may act as an absolute bar or taboo on non-cooperative behaviour (as appears to be the case in many instances in Japan). Or, conversely, the cultural pressures towards accumulation may privilege the pursuit of monetary benefits. It is through such cultural pressures that the costs and benefits of incentive structures enter into actors' decision-making. This is why social capital can work; through engaging with the social relations that constitute a local culture, it can shape incentive structures that, in turn, influence actors' behaviour.

What the social-capital literature emphasizes is the importance of institutional design in shaping incentive structures and determining the nature of co-operation. As E. Ostrom (1995: 127) says 'crafting institutions . . . is one form of investing in social capital'. By implication, it suggests the importance of institutional redesign in altering these incentive structures and changing the nature of collective action. Institutional redesign means a package of measures that will shape the ways that actors within a community will interact over the medium to long term, including the organizational matters, such as:

- nature of arenas for interaction
- role of actors within interactions
- the assignment of rights and duties to actors, and

- the rules of interaction, including monitoring and enforcement, but also norms and routine practice of interaction
- the language of interaction
- the handling of conflicts of values
- the building of agreement, compromises, and trade-offs, and
- the presentation and self-presentation of actors and the interaction process.

In detailing these kinds of changes, it becomes apparent that discursive strategies will be an essential element of creating social capital and changing incentive structures to foster a sense of common interest and hence co-operation. For the last four elements listed are all essentially discursive strategies. They all concern the way in which the communicative aspects of interaction and common identity are handled. But, if such discursive strategies are to build social capital, there are two requirements. First, they need to be relevant to the specific action situation involved. An abstract language of common interest will not be effective; a specific language that speaks to the incentive structures facing actors in that situation may be (Sommer 2000). This in turn requires that the specific community involved is clearly specified. Deliberative democracy may refer to citizens and the community in general terms. The more detailed work of institutional design cannot afford this. The community, the social capital, and the language all have to be specified in relation to each other, and may take a variety of forms: a community of place, identity, or interest, for example (Duane 1997). But specified it must be.

Second, the strategies need to become norms and routines (E. Ostrom 1992). It is not enough for them to be used on occasion and in specific, separated situations; for example, an ad hoc and one-off exercise to develop a certain strategy. They have to become embedded in a continuing institution in which members of a self-identified community can engage in regular contact and develop a sense of reciprocity and mutuality. A medium- to long-term commitment is required for such institutional change; Innes and Booher (1996) recommend years of effort. The community needs an organization basis for mutual engagement and a reason for such engagement. The rules and norms of that institution need, in Ostrom's words, to be crafted and the discursive strategies for collaboration to be embedded in those rules and norms.

Healey's (1996: 229) view that having 'thus generated a knowledgeable consensus around a particular storyline, the task of consolidating the discourse and developing its implications can then proceed' is an ill-founded one. It is insufficient for a collaborative exercise to have generated a strategy that parties feel they 'own' but is then implemented elsewhere. Deliberation cannot be a stage in a larger process of planning. If it is to fulfil its theorists' hopes, it must be a more extensive process. It must become part of the culture of a particular institution, which means, in practice, a planning organization that continually engages in deliberation or a community that works together collaboratively on a long-term project. The big questions that hang over such a process are whether the incentive structures can be sufficiently adapted to ensure that the parties to such deliberation or collaboration continue to want this, and whether the planners are willing to give up or modify their current roles, responsibilities, and power.

Furthermore, such strategies have to become nested within a 'discursive net' (Myerson and Rydin 1996*a*) that supports the community, the institution, the social networks, and the social capital. Strategies are specific tactics and courses of action. The discursive net is an aspect of institutions that involves creating a common way of talking to cement the other aspects of institutional relations. It is the link to the broader culture that makes rules and norms effective. Once that way of talking has been firmly established it becomes a short cut to communication, problem-solving, and self-presentation. In this way, discourse is an integral element of social capital. It is also an essential dimension of establishing a 'common future', not so much in terms of discursive construction of that future—which can often merely cloak a lack of common interest—but in terms of enabling actors to redefine their common interest and then move towards achieving those interests. This whole process involves consideration of the broader discourses used to legitimate policy, which is the subject of the next two chapters.

Conclusion

The discussion has highlighted the centrality of dealing with environmental policy issues as based in the institutional arrangements for allocating and using environmental assets and services. It has shown that perceived conflicts cannot readily be reframed. The discursive strategies available for reframing are relatively weak, precisely because they only seek to represent environmental issues and not actually engage with the institutional basis of these conflicts. Attention, therefore, turned to processes of negotiation and bargaining as a means of engaging more effectively with such conflicts of interest. Analysing the discursive dimension of negotiation and bargaining identified the requirements for a complex and repeated shift between a language of conflict—needed to support strong negotiating positions and define distinctive identifies—and of trust—needed to build relationships between parties that can help cement an eventual agreement. These discursive requirements form part of the reason why negotiation and bargaining often fail.

In turn has come a growth of interest in processes of collaboration and deliberation. But again a focus on the discursive dimension can reveal the demands that such processes place on participants, which were shown to be very considerable in the case of deliberation, compounding the difficulties of the collective action problem constraining participation and the interests of professionals involved (discussed further in the next chapter). The analysis provides support for the idea of building social capital as a way of fostering agreement. But, for this to be effective, the emphasis on building new incentive structures has to be balanced with recognition of the discursive strategies needed to build and maintain those structures. That is, the full implications of building institutions, including the discursive dimension, must be recognized.

5

Rationalizing the Environmental Policy Process

It has been emphasized that deliberative processes need a discursive context, a 'discursive net'. Creating a practical version of communicative rationality that is actively used within the policy process would provide this. In effect, what is required is a justification for a specific policy approach or a legitimation of such a policy approach in terms of communicative rationality. Policy approaches are continuously justified, either explicitly or implicitly. The search for legitimation is an inherent part of the policy process. In contemporary society, such legitimation takes the form of the demonstration of rationality. The reasons for this will be explored and a distinction drawn between procedural and substantive rationality. In this chapter, the focus will be on procedural rationality, the norm of a rational policy process. The next chapter will consider the major substantive rationalities, comparing communicative rationality with scientific and economic rationalities. These are then developed further in the context of three policy case studies in Chs. 7, 8, and 9. The discussion of procedural rationality will be set within the institutional context of policy bureaucracies, considering their interests and analysing the discursive dimensions of their work in terms of plan-making, regulation, and public participation. This will contribute towards a reassessment of the potential for communicative rationality becoming a standard rationalization of policy and deliberation becoming a standard approach to practice.

Rationality in the Modern World

> Intepretative approaches, however, afford us a different view of such rational behaviour, in which individuals and organizations are seen as emphasizing symbols of rationality . . . to foster the image of rational behaviour in a world that expects or needs to see it.
>
> (Yanow 1996: 238)

Describing a particular policy approach as rational legitimates that approach and elevates the importance of those promoting it and implicated in its practice. But why should there be this concentration on rationality within the policy world? An answer to this question is to be found in accounts of modernism and modernity (Giddens 1990). The twentieth century was the era of modernity; it was also the era of the modern state and the growth of state bureaucracies, though its roots can be traced back to eighteenth century French enlightenment (Fischer 2000: 16). Such an era is characterized by a belief in the knowledge and practice generated by scientific methods and a confidence in the ability of organizations (including state organizations) to manage and solve problems: 'an unswerving belief in the power of the rational mind's ability to take control of the natural and social worlds' (ibid.). Alongside the growth of trust in such competencies goes a decline in the authority of traditional institutions such as the aristocracy, the major religious establishments, and patriarchal family structures. Such institutions commanded respect and compliance through a set of values that emphasized hierarchy and deference. Modern institutions, by contrast, work through justifying the rationality and demonstrable competence of their methods. And in democratic regimes, such institutions often have to hold themselves accountable for lapses in rationality and competence.

But analysts of modernity have also been telling us, as we move into the twenty-first century, that we are in postmodern times or, at least, in a distinctive era of late modernity (Harvey 1989). Here many of the assumptions of modernity are being questioned, precisely because of demonstrable lapses in rationality and competence. The institutions are, therefore, being challenged. This applies to state organizations—both elected and appointed—as well as to professional bodies and private-sector organizations such as major corporations. Beck (1992) in particular points to an area of challenge surrounding the inability to manage the risks of late modernity and specifically environmental risks. Within such a society, the centrality of justification to policy language is heightened. There is a broad sense in which *all* language can be said to be about justification (Perelman 1982; Burke 1969; Farrell 1993). Every statement implies a point of view and is seeking to persuade its audience of its relevance and even truth. Stances within a formal debate or informal household row are clearly of this type. But consider other types of communicative interaction: descriptions make a claim to factual accuracy; commands make a claim to authority commanding obedience; questions make a claim to reasonable enquiry and justify the expectation of a response.

In policy communications the centrality of justification is more clearly apparent. Some documents emanating from policy-makers seek to justify a particular stance; some set out alternatives but usually with a preferred option. Often policy-makers are accused of prevarication and a lack of decisiveness if they do not clearly state and argue for their policy stance. Some documents may present themselves as analyses of a problem or a more descriptive, factual account but they do so with an air of authority. And central to all these claims is the notion of rationality. Two aspects of rationality should be distinguished:

procedural and substantive. The former concerns the way in which the policy process is conducted and is discussed in this chapter; the latter focuses more on the goals of that process and the content of policy, rather than how they are achieved, and is the focus of Ch. 6. This distinction is not a completely watertight one. Often assumptions about the content of policy imply assumptions about the way in which policy is carried out. The idea that environmental policy should be based on scientific knowledge and judgements implies that scientists and scientific institutions will be given a privileged role within the policy process. Similarly with communicative rationality, the involvement of a range of actors within the policy process—which is an aspect of policy procedure—is seen as a substantive policy goal; this is the case with much Local Agenda 21 work (reviewed in Ch. 9). Nevertheless, there is a difference in terms of whether a particular justification is primarily concerned with process or substance and this will be used to structure the discussion.

A Rational Policy Process

> A bad condition does not become a problem until people see it as amenable to human control.
>
> (D. Stone 1989: 299)

We prefer our policy-makers to act 'in the public interest'. We may expect them to be swayed by vested interests (even hope for it, if we are the vested interest). We may be disappointed to find they are corrupt and operating in their own pecuniary interests. But the justification for public-sector activity is that outcomes should be better than in its absence. And a key element in producing better outcomes is having a better process for getting to those outcomes. Hence the rationality of the policy process itself is seen as legitimating the activities of the public sector. The belief in the ability of bureaucracies to pursue strategies and routines that are imbued with rationality, resulting in optimal outcomes, has its roots in the very establishment of bureaucracies as a superior form for the state.

Such procedural rationality has a very specific discursive construction, which is summarized in Fig. 5.1, and Fig. 5.2 provides some selected quotes from a publication by the British Department of the Environment (now the Department of Environment, Food, and Rural Affairs), *Policy Appraisal and the Environment* (DoE 1991). This document is a guide for civil service administrators and takes the form of an advice manual, often couched in the second person ('you') with recommendations in bullet points and substantial technical appendices. It includes case studies, summary descriptions of techniques, and even highlighted examples of phrases, sentences, and jargon to incorporate in policy documentation.

Metaphor	Instructions or recipes
Synecdoche	Methodology stands for policy process as a whole; categorization for analysis
Metonymy	Diagrammatic representations such as flow charts
Ethos	Controller
Closure	It is all planned

Fig. 5.1. Rhetorical tropes in the rational policy process discourse

The central theme of this guide is the need to integrate . . . (p. 14)

Proper consideration of these [environmental] effects will improve the quality of policy-making. (p. 1)

the ultimate objectives . . . should be distinguished from the interim objectives . . . (p. 2)

As a first step in the preparation of the guide, the Department of the Environment . . . (p. v)

the key steps in systematically developing and appraising a policy are shown in Figure 1.1. (p. 2)

It is possible for anyone, not just a specialist, to make a first attempt to list the environmental impacts of a policy. (p. 8)

A number of approaches may be used to organise your thinking. (p. 8)

Any form of quantification is likely to provide a better basis for decisions. (p. 10)

use expert advice to identify and quantify significant effects; (p. 13)

In all their decisions ministers need to weigh up costs and benefits. (p. 19)

Fig. 5.2. Key quotes from *Policy Appraisal and the Environment* (DoE 1991)

The key distinctive feature of rational policy process discourse is the assumed benefit of categorizing, listing, and breaking up the policy process into steps and stages, which is most apparent in the metaphorical language of the instruction manual or recipe book that is used to describe and promote the optimal policy process. In the case of *Policy Appraisal and the Environment* it is even apparent in the description of how the manual was prepared: 'As a first step . . .'. Most significantly, it is also apparent in the various diagrammatic representations that are used to capture the essence of rational policy. Indeed these representations, often in flow-chart form, come to stand for the policy process itself and are often sufficient to convey the message of a document, without recourse to the text. Breaking the policy process down into steps, stages, and tiers carries with it a particular conceptualization of policy. It fits with the stages model of policy, critiqued in Sabatier (1999) but also recognized as nevertheless highly durable. In this stages model, formulation is separated from implementation, ends from means, and ultimate goals from intermediate ones. The messy interconnection of formulation and implementation and the repeated redefinition of policy goals that Barrett and Fudge (1981) identified are ignored in favour of a sanitized version of how policy should operate. Furthermore, actors are allocated to roles within these stages, so that particular realms of work are created, particular specialisms and tasks justified.

But while the detailed construction of the discourse emphasizes compartmentalization, the overall claim remains one of policy integration (see the first quote in Fig. 5.2). The rational policy process discourse remains justified in terms of a synoptic ideal of comprehensive data collection and analysis, partly achieved by the claims to comprehensiveness involved in listing, taxonomies, and other proposed analytic techniques. But the role of ethos is also central here; the ethos of the policy analyst being in control of the overall process dominates the discourse. The step-by-step methodologies become a means of maintaining control through managerial modes. Issues of expertise are handled through bringing in outside advice, particularly from scientists, but outside expertise never undermines the advice that the controllers of the policy process themselves give. Rather they become part of a cascade of advice: manuals advise bureaucrats; they take advice from other experts; and they in turn advise politicians. This cascade also erects a wall between political decisions and advice or expertise. It helps present the policy official in a neutral role.

The whole discourse is a very positive one, emphasizing success both actual (through case studies) and potential (through the application of advised methodologies). In this way, the problems of the rational comprehensive model that have been identified by commentators—such as the impossibility of comprehensive data collection, the tendency towards satisficing rather than optimizing, and the problem of incommensurate preference orderings (see Rydin 1998*c*: ch. 2)—are swept aside. The proposed techniques will produce optimal policy outcomes, particularly where they result in quantification. Similarly the messy interaction of interests within policy is ignored; methodology overcomes power.

Not only does this discourse emphasize success, it has been a very successful one. Its ability to justify the role of policy officials has ensured it a continuing life within policy documentation and daily policy practice. Despite criticisms, it remains a benchmark for the operation of the policy process and becomes the mode of expression for the burgeoning applications of environmental assessment, management, monitoring, and auditing. As will be seen below, its existence can become a means of subsuming other discourses; the presumed rationality of the policy process is one of the most significant 'taken-for-granteds' within the process and a major claim that policy officials can exercise. First, though, it is necessary to ask how this discourse relates to the incentives that bureaucrats face within the policy process.

Bureaucratic Interests

The discourse of the rational policy process presents the bureaucrat as disinterested, without particular interests to pursue. Bureaucrats may be vulnerable to the influence of others and, as will be explored in the next chapter, different substantive rationality discourses can be used to justify the involvement or influence of specific interests. However, the above discussion raises the question of the nature of bureaucrats' own interests and how the discourse of a rationality policy process relates to those interests. As the institutionalist approach emphasizes, a focus on the norms of policy practice need not preclude examination of the interest base of that practice.

The traditional sociological literature on bureaucrats has assumed that they pursue status through career advancement and that this involves following the requirements of their particular organization to ensure promotion and salary increments (MacDonald 1995). According to this line of reasoning, bureaucrats can be incentivized by creating organizations that allow for promotion and assure the credentials of status to bureaucrats who achieve promotion. Such an approach fitted with the Taylorist forms of office organization prevalent during most of the twentieth century and could be applied to both private and public sectors. While status and promotion remain significant incentives for any worker, they are insufficient to explain the full picture of bureaucrats' involvement with their work and their behaviour in relation to work practices.

A rather different account of bureaucratic interests has developed within political science focusing on how the bureaucrat could shape the organization to her interests, rather than the other way round. This involves making the bold assumption that the bureaucrat's interests were related to the size of the bureau that she commanded, that is, the number of people working for her and the size of the bureau's budget. As a result it is in bureaucrats' interests to maximize the size of their bureau and this has developed into an 'iron law of bureaucracy'—namely that, in the absence of counter-tendencies, they grow and grow. Intuitively this seemed to fit the twentieth-century growth of organizations, particularly state organizations where there were no constraints arising from the profit motive.

However, Dunleavy (1996) has shown that neither of these accounts fits readily with the changes in public-sector bureaucracies in the late twentieth century, when privatization, hiving off, and other forms of organization change were common. Instead Dunleavy has developed a bureau-shaping analysis based on an institutional rational-choice approach. This argues that there are four reasons why bureaucrats do not budget-maximize. First, maximizing a budget requires collective action and bureaucrats, like environmental protestors, face collective action problems. Second, the relationships between budget size and bureaucrat's utilities vary across budget type and agency type. Third, any incentives for budget maximization only work up to a certain threshold. Fourth, senior bureaucrats are more likely to pursue non-pecuniary utilities, not associated with budget size.

Dunleavy argues that all bureaucrats face a collective action problem in trying to get growth to happen but this is reinforced by dynamics specific to agency type. He identifies eight types of bureau or agency: delivery (of outputs or services), regulatory, transfer (passing subsidy or entitlement to private individuals and firms), contracts (for services or capital projects), control (channelling funds to other public-sector bureaux under supervision), taxing, trading, and servicing (to other public-sector bodies). These different agencies then have different types of budget, which can be analysed in terms of four nested elements:

- core budget spent on the bureau's own activities;
- bureau budget, incorporating the core budget plus any moneys paid out to the private sector;
- programme budget, incorporating the bureau budget plus any moneys that the bureau passes onto other public-sector agencies; and finally,
- super-programme budget, incorporating the programme budget plus any expenditure by other bureaux over which the agency exerts some control, asserts some responsibility, or claims some credit (or incurs some blame).

Most agencies involved in environmental planning are regulatory agencies, in which the core budget takes up a substantial proportion of bureau and programme budgets, with any difference between core and bureau budget only arising from the use of subsidies to back-up regulations. Both the core budget and bureau budget will grow steadily with the overall size of the bureau and the programme budget; there is no super-programme element. In such agencies, there is an incentive to increase core budget, as this is a steady proportion of overall budget. Pennington's (1997) work on the British land-use planning system suggests that there are such internal growth pressures in environmental regulatory agencies but they should be seen in the context of the overall relatively small size of such agencies. The scale of such agencies in general seems to be small. Figures for the UK central state in 1987–8 show that regulatory agencies had an average budget of £77 million and staff corps of 3,800 compared to an overall average of £250 million and 13,400 staff (calculated from Dunleavy 1996: 188).

To consider other dynamics, it is necessary to specify a little more fully the motivation and incentives affecting bureaucrats. While Dunleavy does not seek to provide a psychological model of bureaucrats' behaviour (and indeed is critical of attempts to do so), he does review the organizational sociology literature to cull a list of positively and negatively featured values of work, which will influence bureaucrats' decision-making. These are summarized in Fig. 5.3. From this Dunleavy (ibid. 202) argues that bureaucrats will seek to shape their work and bureau so as to enhance positively valued features and hence contain bureau growth: 'Rational officials want to work in small, élite, collegial

Positively valued features	**Negatively valued features**
Staff functions: Innovative work; longer time horizons; broad scope of concerns; developmental rhythm; high level of discretion; low level of public visibility	*Line functions:* Routine work; short-time horizons; narrow scope; repetitive rhythm; little discretion; high level of grass-roots/public visibility
Collegial atmosphere: Small work unit; restricted hierarchy; lots of élite personnel; co-operative work patterns; congenial personal relations	*Corporate atmosphere*: Large units; extended hierarchy; many non-élite personnel; coercion and resistance in work patterns; conflictual personal relations
Central location: Close to political power centres; metropolitan; high-status social contacts	*Peripheral location*: Remote from political power; provincial location; remote from high-status contacts

Fig. 5.3. Positively and negatively valued features of bureaucratic work (from Dunleavy 1996: 202)

bureaux close to political power centres. They do not want to head up heavily staffed, large budget but routine, conflictual and low-status agencies'. This will particularly be the case for more senior officials.

It is clear that this dynamic is as likely to apply to senior officers involved in environmental planning as elsewhere. Indeed, it is probably a particularly applicable argument as the environmental area involves many with professional qualifications (such as planners, environmental health officers, etc.) who will aspire to and, indeed, expect some of the positively valued features listed in Fig. 5.3 as attributes of their everyday work experience. Thus environmental agencies, particularly those in professionalized areas of work, are likely to be shaped by their bureaucrats so as to maintain a relatively small budget and size, and to privilege policy work with the above positively valued features.

This institutional rational-choice analysis provides an insightful account of why environmental bureaux are as they are. But, for a complete account, the role of language in these dynamics should be acknowledged. What is notable about the list in Fig. 5.3 is that, while they are presented by Dunleavy as factors that contribute to a utility function for bureaucrats, they almost all deal with the interrelation between bureaucrats and with others within the work environment. Thus they depend on communication between people. At minimum, it can be said that language is a necessary feature of the expression of these attributes. Language is required to give effect to congenial working relationships. But, as argued throughout this book, language is not a neutral medium. The discursive resources available within an organization open up and close off opportunities for communication.

The issue is closely related to that of the norms and values of an organization. The attributes included in Dunleavy's table could, in most cases, also be characterized under the heading of organizational norms and values. Characterizing them in this way emphasizes their long-term nature. Norms and values are not just a one-off expression of a particular feature but have to be embedded in an organization so that they continue to exercise influence over the longer term and without conscious individual action to reproduce them. Language and its construction in terms of organizational discourses play an important role in embedding and reproducing norms and values and ensuring that the patterns of attributes have a more than transitory life—a similar point has been made in relation to social capital. To explore these issues in more detail, different aspects of planning work will now be examined.

Plan-Making

Policy-making and the development of plans, programmes, and proposals (here all handled under the heading of 'plans') is work that carries high status and little budget. The previous section has already established that such work is precisely the kind that bureaucrats will strive to undertake and that their preference will result in relatively small policy units that do not grow at any substantial rate. The preference for this type of policy or plan work is readily found in most environmental policy organizations. In land-use planning, development control (the regulatory function) has long been seen as the 'Cinderella' to development planning; in environmental regulation, routine pollution control is seen as less attractive than policy-making, and often hived off to lower-level organizations. The bureau-shaping model appears to have resonance here. But what role does the language of plan-making play in everyday practice in bureaux? The language of plans arises from a number of different dynamics, external and internal to the plan itself.

First, there are the implications of the multiple roles that plans are required to fulfil. In a study based on the British land-use planning system, Healey *et al.* (1988: 193) identify six roles that plans can play. These amount to informing decision-making (including regulatory) both within the organization and outside, and establishing a framework for the investment and allocation of resources by various bodies, and are classic roles for plans that are justified by the rational policy process discourse. They also shape the linguistic form of plans. Both regulation and resource allocation are tasks that are performed according to certain organizational norms and standard working practices; they will be discussed further below in relation to regulation. Therefore the language of plans has to mesh with the language of regulation or allocation, as performed elsewhere within the organization or by another organization. The linguistic content of a plan or policy becomes a resource to justify regulatory or investment decisions and, to be useful in this role, the language has to carry weight as a justification in these arenas.

It is worth remembering, though, that the extent to which regulatory decision-making is tied to the regulatory framework, and vice versa, is partly a reflection of the resources for implementation controlled by the regulatory bureau. Where the majority of resources are held outside the regulatory bureau—as with British land-use planning—then the amount of discretion allowed to regulators has to be considerable. On the other hand, where the same authority controls resources for implementation and regulatory decision-making—as in much Scandinavian land-use planning, for example—then regulatory discretion can and will be reduced. It is to be expected that this difference will also be reflected in the language of the policy process. Where there is scope for more discretion, a less formalistic mode of expression may be used, which gives scope to uncertainty and ambiguity and offers more opportunities for alternative interpretation. However, where discretion is less, the language of the framework will be more certain, linear in its reasoning, and closed in terms of interpretation opportunities.

But plans do more than provide a framework for regulation. In their study, Healey *et al.* show how plans also play a role in mediating conflicts between actors and establishing an acceptable compromise. The very act of creating the plan will involve negotiation and bargaining and, therefore, reflect the language of handling conflicts discussed in Ch. 4. More specifically it has to give expression to the settled outcome. This will involve signalling the ways in which the plan-making process has considered and made allowance for specific interests. It will also need to justify its reasoning and any trade-offs made, even if only by invoking the higher authority of the plan-maker. It may also choose to justify why particular courses of action were not taken. Thus the language of the plan has to reflect the conflicts and compromises that went into its creation as well as the demands of regulatory decision-making.

Second, the plan is not just a historic record of the pressures on its creation but must continue to address a variety of audiences. And here there tends to be a difference between commentators on how plans discursively tackle the demand of multiple audiences. Some, such as Healey, emphasize the potential for a unifying or at least dominant discourse within the plan arising from the process of plan-creation. While she writes of 'what the text of a plan could mean in the flow of events and relations among the interests who played the game of plan construction and the characters who constituted its drama(s)' (1993: 84), and points out that 'Potentially several discourses may coexist within a single plan' (ibid. 85), she holds out the prospect of a unified or dominant storyline (1996: 229). Others, however, see fragmentation and contradiction as likely to remain embedded in the plan. Hillier (1993: 90), for example, emphasizes that 'Planners use different language patterns and terms, and even tell different stories according to the audience they are addressing', and writes of multiple stories running through a planning document resulting in 'tortuous plannerspeak' (1997: 31). Similarly, Mazza (1986) has pointed out that internal consistency is an unlikely property of plans, though he sees this as arising from the interactive contexts in which they are produced rather than a necessary feature of their role as documents to be read.

There are good reasons for assuming that plans are more likely to contain multiple storylines and be internally inconsistent. First, it is advantageous to plan-makers to maintain a degree of ambiguity within the plan in order to satisfy their ongoing constituencies, as Hillier and Mazza clearly foresee. Second, it can be a discursive resource of an internally contradictory plan that a number of different bases for a variety of regulatory decisions can be found within the one document. When any form of legal or procedural challenge to a plan or decision takes place, the various parties to the challenge will search for these multiple lines of reasoning within the document to justify their own particular position. Thus in an appeal within the land-use planning system, planning documents will be scoured for words that allow different interpretations to suit the appellant.

But this need not involve discrete elements within a plan, certain sections justifying one line of action, other sections justifying others. This may be the case but it is one of the features of language that a given phrase, sentence, or section can be interpreted in a number of ways. This essential ambiguity of language is one of its most important features as a resource and this is a third reason why a plan will be a multiple rather than a unified discourse. Thus different parties will not search a document only for different sections to justify their actions, but for different interpretations of the same section. In her empirical study of development plans, Healey (1993: 91, 99) writes of how one plan was dominated by its role in managing ambiguity and another was characterized by considerable ambiguity in content and style.

Despite such pressures for diversity, multiplicity, and ambiguity, a plan must hang together to some extent. The various sections, different lines of reasoning, and different possibilities have to form a document. The document need not, and is unlikely to be, coherent and consistent throughout but it should still satisfy the justification provided by the rational policy process discourse and, as van Woerkum (2002) suggests plan-makers will wish to infer that the document is unified, coherent, and even logical. Therefore there are limits to the incoherence and inconsistency of such a document, limits created in terms of style and content. Any particular document will also have to conform to the stylistic requirements of its particular genre. Bramley (1985) provides a useful analysis of the linguistic conventions of the British land-use planning document known as the 'written statement', the textual element of a spatial development plan; this is summarized in Fig. 5.4.

The problem of satisfying multiple roles and constituencies through a degree of ambiguity and inconstancy while yet maintaining some coherence is also solved by the framing of the policy issue. Framing has already been discussed in Ch. 4 in the context of Schön and Rein's work. There the emphasis was on reframing a policy issue so that conflicts became less troublesome. Here, however, the point is that the way in which the policy issue is framed encourages a

Linguistic category	Analysis of Written Statement
Graphetics	No colour; no right justification; quas-legal purpose; 'report' rather than 'work' denoting shorter life-span
Graphology	Use of numbering system stresses complexity of issues and expertise of author; capitalization suggests announcement takes precedence over explanation or justification
Grammar	Substantial repetition; ellipsis avoided; limited use of pronouns and adverbial contrasts (such as 'however'); all arising from quasi-legal purpose
Vocabulary	1. Technical terms 2. Everyday terms used slightly differently 3. Lay terms Majority of terms tend towards 1 or 2.
Semantics	Not covered by Bramley's article

Fig. 5.4. Bramley's linguistic analysis of the genre of 'written statements'

dominant reading but allows various lines of reasoning and interpretation to be pursued at the same time. Healey's work has shown how planning documents frame development and environmental issues in particular ways. Writing of a specific plan, she states 'The main communicative work of the plan is to *structure* the agenda of debate within the ongoing conversation between planners and developers, and *impose* the strategic agreement on everyone else' (1993: 96, stress in the original). Another plan is analysed within the same article as justifying a particular approach, which is another form of framing (ibid. 99). Thus the language of plan-making is a central element in creating plans, enabling them to speak to multiple audiences, allowing them to fulfil their policy roles, and yet maintain a relatively distinct policy agenda.

Regulatory Practice

The standard model of regulatory practice, reinforced by the rational policy process discourse, is that once policy has been formulated it is implemented and regulatory practice follows the guidelines set down in policy documents. However, research on the implementation of policy and, in particular, regulatory practice has repeatedly shown that this model does not describe real-life practice. First, the apparent 'stages' of formulation and implementation are not distinct but merge into one another (De Leon 1999). Barratt and Fudge (1981), some two decades ago, showed how policy practice moves back and forth between the formulation and implementation of policy, with many details of policy being revised during the experience of implementation, including even the goals of policy. Thus regulatory practice is not distinct from the policy framework for regulation but itself helps to shape that framework. This is another reason why the language of the policy framework meshes, to some extent, with the language of regulation.

In a study of the meaning given to environmental issues in British development plans and development control decisions (specifically appeal decisions), it was found that 'the network of meanings around these uses of "environment" in development plans is much richer than is possible within the rules and occasions of decision letters' (Myerson and Rydin 1994: 448). This meant that the development plans could draw on a discourse of the sublime, the romantic discourse of wonder, and personal engagement with nature. In the regulatory discourse, however, 'where the phrases of "environment" occur, the pattern of usage was remarkably consistent' (ibid. 444). In regulatory discourses such consistency is a required feature, arising from the legal authorization that regulatory decisions convey. By contrast, in development plans a different form of authorization is at work, a broader political or cultural authorization about the legitimacy of the plan-making process. This will involve taking account of a broader range of issues and interests, referenced through a broader range of discursive constructions as noted above.

Second, the decision-making involved in regulation does not simply follow the established framework. Rather there is always a level of discretion involved in regulation and, sometimes, that discretion can be considerable (Lipsky 1980; Gouldson and Murphy 1998). Greater bureaucrat autonomy is generally associated with greater discretion. This does not mean, however, that the strict guidelines set down in a policy framework are replaced by complete anarchy in terms of regulatory decisions. Discretion rather implies some degree of choice by the bureaucrat in how to apply and interpret the guidelines. In exercising this choice, bureaucrats will undoubtedly be influenced by their own values and some may even have their own pecuniary interests at heart. Leaving such instances of corruption aside, in most regulatory discretion it is the norms and standard working practices of specific groups of bureaucrats that will have the most influence.

One way in which these common languages and norms of working practice can be summarized is in terms of 'styles' of regulation, following Vogel (1986). Applying Vogel's notion of style to environmental regulation, Gouldson and Murphy (1998: 49, 65) provide a detailed account of how style can vary with the stage of implementation and the overall approach adopted in two examples of environmental planning practice: the mandatory regulations of the EU's integrated pollution control and prevention directive and the voluntary environmental regulation of environmental management and audit systems. These regulatory styles provide a descriptive typology of regulatory behaviour. They arise from the inevitably discretionary element within regulation and represent an attempt to handle the more complex interrelationship of regulation and plan-making that occurs in practice. Each style represents a particular package of institutional arrangements, working practices, norms, and values, and the expression of all these facets through discourse.

The 'styles' literature may describe sets of norms operating in particular organizations, but it is also important to understand why such sets of norms or styles arise within organizations. According to most sociological accounts of professions and organizations, norms and practices are a way of creating a distinct working identity, thereby cementing relationships among a group of workers and separating it from other groups. Formal qualifications and entry requirements may also be important in these sociological processes, but they are reinforced by the daily exercise of standard practices and the routine expression of values and norms of the group (Johnson 1972; MacDonald 1995). Such sociological dynamics are clearly important but there are also other incentives to adopting standard practices and norms; these relate to the time and costs of working practices. Here standardization of communication can be seen as a short cut, reducing the effort of bureaucratic work. At the same time they can enhance the positively valued features of bureaucrats' work. The sense of discretion, of specialist work practices, and of a collegial atmosphere can all be enhanced by the use of specialist language. It is not just peer pressure but the joint benefits appreciated by each bureaucrat that encourages the use of such language.

The application of the package of standard norms, working practices, and modes of communication within regulation has significant consequences. In their study of negotiating planning-gain agreements for environmental goals, Whatmore and Boucher (1993) found that one particular discourse of nature dominated the construction of this particular regulatory device. While they identify three possible discourses of nature—conservation, commodity, and ecology—they found that the commodity narrative was strongly associated with the practice of environmental planning gain. But they also found that this practice was not widely used; environmental planning gain had not become routinely institutionalized. They analysed this in terms of the discursive tensions involved in the commodity narrative, which itself was weakly institutionalized within British planning and vulnerable to discursive challenge. Such challenges came from two sources: the professionalized community of planning, which has institutionalized the conservation narrative, and the environmental movement, which has promoted the ecological narrative. These competing narratives were able to challenge the commodity narrative on its interpretation of both 'environment' and 'gain', partly because of the institutionalized regulatory practice of planners.

Of course, in some systems there is considerably less bureaucratic autonomy. For example, the US system is generally cited as one where discretion largely operates at the level of the judiciary rather than the regulator. But this does not mean that the importance of working practices and norms within the regulatory bureaucracy is reduced. Rather they have to be shaped to take account of discretion operating elsewhere. In a legalistic system, such as the US, the importance of decision-making in the courts will shape the language of policy documents and of everyday decision-making. Thus Killingsworth and Palmer (1992) point out how the threat of legal challenge resulted in environmental impact statements becoming incredibly lengthy so that they could contain all possible material anticipated as necessary to counter arguments put forward in legal challenges. The discursive construction of regulation remained a significant response to and influence on the nature of the environmental policy process.

Public Participation

One theme that has run throughout this book concerns the potential for a more participatory, deliberative, communicative approach based on a careful examination of the interests involved, and the discursive challenge such an approach poses. The final section of this chapter, therefore, considers how bureaucrats' interests may be affected by political demands for more participation and public involvement in decision-making. In the previous chapter it was suggested that one of the problems facing the dissemination of deliberative approaches was the lack of policy champions. This chapter therefore tries to answer the question: How can it be in bureaucrats' interests to involve the public?

Recalling the discussion of Ch. 4, public participation cannot be taken for granted. The dynamics of the collective action problem discussed there mean that the public is often reluctant to be involved because of the imbalance between the benefits and costs of involvement. Furthermore, such participation that does result is often highly skewed, in a way that benefits middle-class protestors and those with a spatial patch to defend. Therefore, extending participation is a difficult task. It has been suggested that developing forms of social capital may be a way to overcome the collective action problem; that again can require considerable effort.

Handling the collective action problem might therefore require specific expertise and resources, which can become a specialist area of expertise for bureaucrats and carry the benefits associated with bureau-shaping. It may be that it is in certain bureaucrats' interests to develop a unit in which policies for public participation are developed. This work carries with it many of the positively valued features identified by Dunleavy, although it does also involve exposure to public visibility and grassroots contacts. For some bureaucrats, though, that will be compatible with their personal, political, and professional identity—the grassroots professional. That such a new breed exists is evidenced by the way that some bureaucrats and planning professionals are using the communicative rationality discourse (discussed further in Ch. 6). But in focusing on the role of bureaucrats in opening up the policy process to outside groups, it can be seen that having influence over access to decision-making could be a high-status activity, particularly where there are a large number of actors already influencing policy. Where policy is opened up to a wide range of influence, then the control of the bureaucrat over the policy itself can be limited. Shifting expertise and influence to the structuring of the policy situation rather than the content of policy would be a rational shift for those bureaucrats who see this opportunity and are able to take advantage of it.

However, the status of expertise is always limited by its effectiveness, and the stubborn nature of the collective action problem may itself limit the opportunities for a professional who defines her expertise in terms of greater participation. Furthermore, this discussion of the potential for change should not detract attention from the fact that most routine public participation practice is rather mundane. It consists of poorly attended exhibitions and consultation activities alternating with vociferous protests, which bureaucrats find themselves unable to respond to, within the limits of their work. This is both frustrating for bureaucrats and threatening, since it suggests their inadequacy to generate acceptable policy outcomes; such protests can threaten the legitimacy of planning practice. Therefore, it is in bureaucrats' interests to shape public participation, so that the demands raised have at least the potential for being met within the current structures of planning practice. In this way, bureaucrats manage participation events and arenas, in order to diffuse protest and channel it in ways that fit with norms and standard working practices.

Van Woerkum (2001) describes this as 'conquering' the possibilities offered by a culture of orality planning, a culture that offers scope for public involvement but also threatens established modes of planning practice. He describes this culture of orality as time-bound, slow, auditory, and memory-dependent, but also interactive, creative, and flexible. By contrast, he argues that most planners are bound up in a culture of literacy in which they privilege the 'plan', imbuing it with symbolic qualities. While citizens see the plan only as an input to further discussion and decision-making, planners see the plan as an output. These contrasts can become a conflict of cultures and this poses three choices for planners. They can conquer orality and defend the plan (as intimated above), they can try and cope with orality but no more (a low-status option), or they can try work with orality and use it. However, van Woerkum is able to provide much more detail on how planners conquer orality than on how they might use it.

One classic way of managing public involvement has been through the use of forums such as public inquiries. Public inquiries have been one of the growth areas of environmental planning practice. Not only are they commonly used to present debate about environmental issues in public, but they are also often incredibly lengthy, taking up considerable bureaucratic time. They have a number of advantages from a bureaucratic perspective. As indicated above, they can channel participation into a form that fits with the rationality of the policy process, allowing for the discussion of compromises and diffusing protest that challenges the legitimacy of environmental decision-making. Of course, public protests will sometimes try to subvert these arenas as with the classic case of anti-motorway protests and their disruption of public inquiries (Wall 1999). They also involve relatively high-status work for bureaucrats, making public appearances and sometimes working alongside very high-status professionals such as barristers and private-sector consultants. Finally, they are an area where bureaucratic growth is possible and indeed sanctioned.

The discursive dimension of public inquiries has been studied alongside their organizational features. Most research highlights the way in which the inquiries result in an inadequate treatment of the policy issue and favour some interests at the expense of others. This has been a particular emphasis in studies on the 'nuclear' inquiries, into nuclear power stations and waste reprocessing facilities (Ince 1984; Armstrong 1985; Kemp 1985; O'Riordan, Kemp, and Purdue 1988). Almost all of these studies have noted that, despite heroic efforts from local and environmental groups, the nature of the discussion at the inquiry meant that the odds were heavily stacked against them. Part of this bias arises from the simple issue of resources. Commenting on the Sizewell Inquiry, Ince (1984: 6) says:

> The inquiry could easily have turned into a fixture more reminiscent of the lion's meetings with Daniel than the gladiatorial combat which it is popularly held to resemble. However . . . the objectors did well at the inquiry out of all proportions to the resources available to them. But I do not think that this means that the gross inequality of resources at the Sizewell inquiry can be regarded as acceptable.

Armstrong (1985: 133) agrees with this conclusion in her assessment of the proceedings and it is also echoed in O'Riordan, Kemp, and Purdue's (1988) study.

Another aspect of the bias arises from the kinds of issues that are allowed into the discussion at such inquiries. The nuclear inquiries have often been faced with protestors seeking to widen the terms of the inquiry in order to admit discussion of national energy policy, national defence policy, the need for the development or economic viability, issues that have been considered outside the scope of the inquiries' proceedings. While procedural rules can limit which issues are discussed, this is also a more general feature of the inquiry format. Even though O'Riordan *et al.* found that the Sizewell inquiry suffered from considerable information overload, they listed many areas where discussion had been insufficient. They analyse this as partly arising from the attempt to deal with national policy issues at a site-specific inquiry. But it is the case that any format will impose some constraints on the scope and shape of discussion. Comprehensive coverage is not possible; every format for discussion will structure the discussion in certain directions.

Ince's comment, quoted above, is interesting for it recognizes that, while issues such as resource inequality are clearly important, much more is going on at an inquiry. There were communicative possibilities that the protestors were able to take advantage of despite their limited resources and there is the image of 'the gladiatorial combat' to consider also. Such inquiries have a very particular adversarial structure based on the presentation of evidence and subsequent cross-examination, sometimes with questioning by the chair or one of her panel (if she has one). This is clearly a quasi-legal structure and, particularly where legal professionals are involved, can lend itself to highly legalist form of argumentation and even the use of legal jargon. Such discursive practices are one of the main foci of criticism by community groups, who find them disempowering and exclusionary (see Armstrong 1985).

Furthermore, the discussion at inquiries encourages linear argumentation and an emphasis on specific points of logic within each party's case. The discussion can become very defensive, as presenters of evidence seek to make their points in the face of cross-examination. The larger picture can be lost in the detail of such argumentation. Issues of values, feelings, and emotion are considered as inadmissible, since they are not open to cross-examination or support through logical arguments. Thus, O'Riordan, Kemp, and Purdue (1988: 405–9) call for a greater variety of formats for discussing issues including preliminary meetings, side meetings, local public meetings, and on-floor hearings.

In their analysis of how different formats for public discussion shape that discussion, Bryson and Crosby (1993) identify three different types of format: forums, arenas, and courts. They see each of these as performing different functions and linking into different dimensions of power; this is summarized in Fig. 5.5. These organizational settings and the type of discussion and communication that occurs in these settings influence the ideas, rules, modes, media, and methods that constitute everyday planning practice. So the design of these settings is centrally important. But it is also necessary to note Bryson and Croby's caveat: where there is no intention to share power on the part of the powerful,

	Forums	**Arenas**	**Courts**
1st dimension of power	Creation and communication of meaning	Policy-making and implementation	Management of residual conflict and enforcement of underlying norms
2nd dimension	Communicative capabilities Interpretative schemes Norms of pragmatic communication	Policy-making and implementation capabilities Methods Agenda Domain	Conflict management Sanctioning capabilities Jurisdiction
3rd dimension	Signification	Domination	Legitimation

Fig. 5.5. Bryson and Crosby's forums, arenas, and courts

then such settings will be structured to maintain existing power relations and not to achieve any mutual goals. The fear of many commentators on highly significant public inquiries is that they have been an exercise in legitimation through obfuscation rather than any real attempt at communication of any kind.

Finally, whatever the nature and extent of the public participation activities, consideration has to be given to the way in which the views raised by outside groups and individuals are treated. Here, the danger is always that the procedural rationality of the policy process will take over. As Healey (1996: 226) points out: 'Such material is translated into, and filtered through, the technical language used by planning analysts and the administrators of the planning system. This will almost immediately reduce a person's speech into a "point", to join other points in a structured analytical framework through which the planners seek to "make sense" of what is going on.' She also points out that most consultation takes place only after the main options for choice have been identified and limited in number. She argues (with others, such as Innes and Booher 1996) for opening up consultation at an earlier stage. But it is one of the ironies of the collective action problem that people are less likely to see it as in their interests to become involved while the policy options remain vague and undefined. Public participation at strategic planning stages has always been less vigorous than on detailed development proposals.

Thus—to reflect back on the question posed at the beginning of this section—it can be in bureaucrats' interest to foster public participation but they are likely to do so only in ways that benefit them. This may include attempts at community empowerment but it may also include abortive consultation exercises and rather unsatisfactory and formalist arenas such as public inquiries. Public participation and professional practice do not necessarily go together. As Healey and Hillier (1996: 172) point out in their study of planning in Swan Valley, Perth: 'As time passed, however . . . officers realized that more participatory forms of planning often take a long time, and that the local community understands and knows far more than the planners originally thought. Tension escalated.' With this context in mind, we can now move on to explore the substantive rationality that might support greater participation and empowerment—communicative rationality—and see how it relates to other substantive rationalities of environmental planning.

Conclusion

This chapter has explored the self-image of the planning process, and of policy processes more generally, as rational. This is embedded in the assumptions and norms of contemporary modernity, and has not yet been shaken by any movements towards postmodernity. The rational policy process discourse is a powerful one, because it has strong connections into the self-interests of policy bureaucrats. It has been shown how this affects all aspects of bureaucrats' work, in plan-making, regulation, and organizing public participation. The discussion of this last point, in particular, has reinforced concerns raised previously about the problems involved in achieving more substantial and meaningful public engagement in the planning process. It remains persistently difficult to put communicative rationality into practice. This raises questions about the nature of communicative rationality as a practical policy discourse, which are explored next.

6

Discourses of Environmental Rationality: Three Rationalities

Environmental planning is, therefore, constituted by a procedural rationality. However, it has been emphasized that policy practice is also shaped and legitimated by substantive rationalities, what V. Ostrom, Feeny, and Picht (1988: 458) call 'different grammars of choice'. Specific instances of environmental policy are justified in many different ways. It is one of the features of the discursive dimension that it allows for considerable creativity in the development of justificatory discourses. But the processes for maintaining and reinforcing discourses, discussed in Ch. 3, suggest that the main discourses used in any specific context will be limited in number. In the case of environmental planning, there appear to be three main sources of rationality that are used to legitimate policy and decisions: scientific, economic, and communicative. These rationalities are, of course, used widely within society to justify all sorts of decisions and actions, but they appear to have particular resonance in the environmental domain. In this chapter these three substantive rationalities are analysed. As well as adding to our knowledge of how environmental planning is legitimated, the analysis enables further reflection of the extent to which a substantive rationality of communicative rationality can legitimate changes in actual planning practice. The chapter concludes with a discussion of the interaction of these substantive rationalities with procedural rationality. In the next three chapters case studies are presented, in which the contextualized analysis of these rationalities, together with procedural rationality, can be extended. The issue of the interrelationship between these three rationalities is explored in Ch. 10.

Scientific Rationality

The reliance on scientific knowledge has become a key element of environmental policy. It is commonplace to find arguments that 'sound science' should be the basis of such policy and, organizationally, scientific advisers are found in relatively central locations within many bodies from the World Bank down to local councils. Our appreciation of the very existence and extent of environmental problems such as global warming, loss of biodiversity, and depletion of the ozone layer arises from the activities of scientists. As Beck (1992) has pointed out, these are 'invisible' problems, which can only be 'seen' through the medium of scientific techniques. Hannigan (1995: 82) states: 'It is rare indeed to find an environmental problem which pops up overnight with no past legacy of scientific observation and debate.' Hence science has a central role to play in specifying the problems and issues of the environmental agenda and, by extension, in identifying the solutions. Scientists are in a position to influence not just the agenda but the evolving institutional arrangements for achieving policy change and, thereby, changing the incentive structures facing actors.

The claims of scientific rationality are so widely accepted within society that it is almost a matter of self-reflection to identify them. Science is accepted as the source of knowledge in our society. It is seen as predominantly objective, neutral, and the route to truth. It enables facts to 'speak for themselves'. This is assumed through the scientific method, firmly based in the use of experiments, which can be replicated with identical results. Experiments allow the falsification of theories and without experimental proof, theories remain provisional. Scientists, therefore, study reality; even allowing for the insights of Schrödinger that the very act of measurement changes the reality being measured, for most science above the atomic level it is accepted that reality is 'out there' to be studied. This acceptance of science as the route to objective knowledge of our work has been identified by Latour (1999) as part of the 'modernist settlement'. This is a crude pen-portrait of the claims of scientific rationality, but one that nevertheless applies in most non-scientific circles and, as Killingsworth and Palmer (1992: 126) point out, when science enters the public realm it becomes 'something other than science as defined by scientific authorities'. It is a view that counterposes expertise to the general public, seen as largely passive and incompetent (Throgmorton 1991).

It is not just scientists and scientific institutions, though, that argue for a prominent role for scientific knowledge in shaping environmental policy. Scientific rationality has also become a mainstay of many environmentalists' arguments. While De Luca (1999: 59) considers that the image events of Greenpeace often refute scientific rationality, others have repeatedly shown how environmentalist groups seek to use this rationality (Hannigan 1995). Through this means they can argue for the pre-eminence of biophysical systems in setting constraints on our actions. According to such arguments, scientific knowledge of these systems reveals the necessary limits to social behaviour, if environmental catastrophe is to be avoided. This can support calls for more environmental protection. It also suggests a definition of sustainable development, in which the environmental or ecological constraints set the context for economic and social considerations, rather than a more radical interaction between the three dimensions being sought. How does scientific rationality work discursively?

As a way of investigating this, a key example of the expression of scientific rationality will be analysed. The chosen example is Tom Lovejoy's Reith

Lecture. Lovejoy is Chief Biodiversity Adviser for the World Bank and Counsellor to the Smithsonian Institution, USA. Along with other luminaries such as Gro Harlem Brundtland, Vandana Shiva, Chris Patten (EU Commissioner), John Browne (Head of BP Amoco), and the Prince of Wales, he was asked to contribute to the BBC's Reith Lectures for 2000 on the theme of sustainable development. As befits his position, his talk focused on threats to biodiversity and presented the scientific viewpoint. A rhetoric line for Lovejoy's speech (and, later, for the Prince of Wales's speech from this series) is presented in Fig. 6.1, along with Fig. 6.2 identifying the key tropes of the rational scientific discourse that Lovejoy draws on.

The rhetoric line of Lovejoy's speech (Fig. 6.1) shows that he does not argue in a simple linear way, points are often disconnected, and there is a lot of movement from specific examples to general points. The overall point that he wants to make is clear though and repeated several times: the centrality of biodiversity to the sustainability agenda and the way that science can provide knowledge about biodiversity. In particular he calls for biodiversity to be the measure of sustainability and, rhetorically, he uses the synecdoche whereby biodiversity, which might be considered part of sustainability, comes to *be* sustainability. The scientific understanding of the environment becomes the defining feature of sustainability, and hence sustainable development. This is an explicit argument conducted through synecdoche and supported by other synecdochic rhetoric, such as where a particular species—the gnatcatcher—is used to stand for the larger ecosystem or a particular site is used to stand for broader habitats. This rhetorical line weaves its way through two different discourses. The first, which is emboldened in Fig. 6.1 and can be seen to dominate, provides the traditional view of science as rational and knowledgeable. But, as an environmentalist, Lovejoy seeks to combine a reliance on scientific rationality with an acknowledgement of his environmental values. He does this through the subsidiary use of a romantic discourse, indicated in Fig. 6.1 by italics. Each of these will be discussed in turn.

The discourse of scientific rationality is a self-confident discourse, particularly when used by an accredited scientist (although one speaker from the floor at Lovejoy's Reith Lecture referred to it more disparagingly as a 'rational smokescreen'). Within this discourse there is an emphasis on knowledge arising from experiments and painstaking study. Metaphorical language is limited to a description of biodiversity as a library, which allows bibliographical work. It is a highly empiricist discourse in which, through the rhetorical device of metonymy, the facts identified by the scientists are taken to stand for the object of study itself. Thus biodiversity loss is 'really a given' and not to know the full extent of biodiversity is 'scandalous'.

The key ethos that is invoked is that of the professionally recognized scientific expert, who is often cited by name: E. O. Wilson, 'the great Harvard biologist' and Ruth Patrick, holder of the US National Medal of Science (although, as a woman, also a 'den mother' and a 'grandmother'). Scientific training is

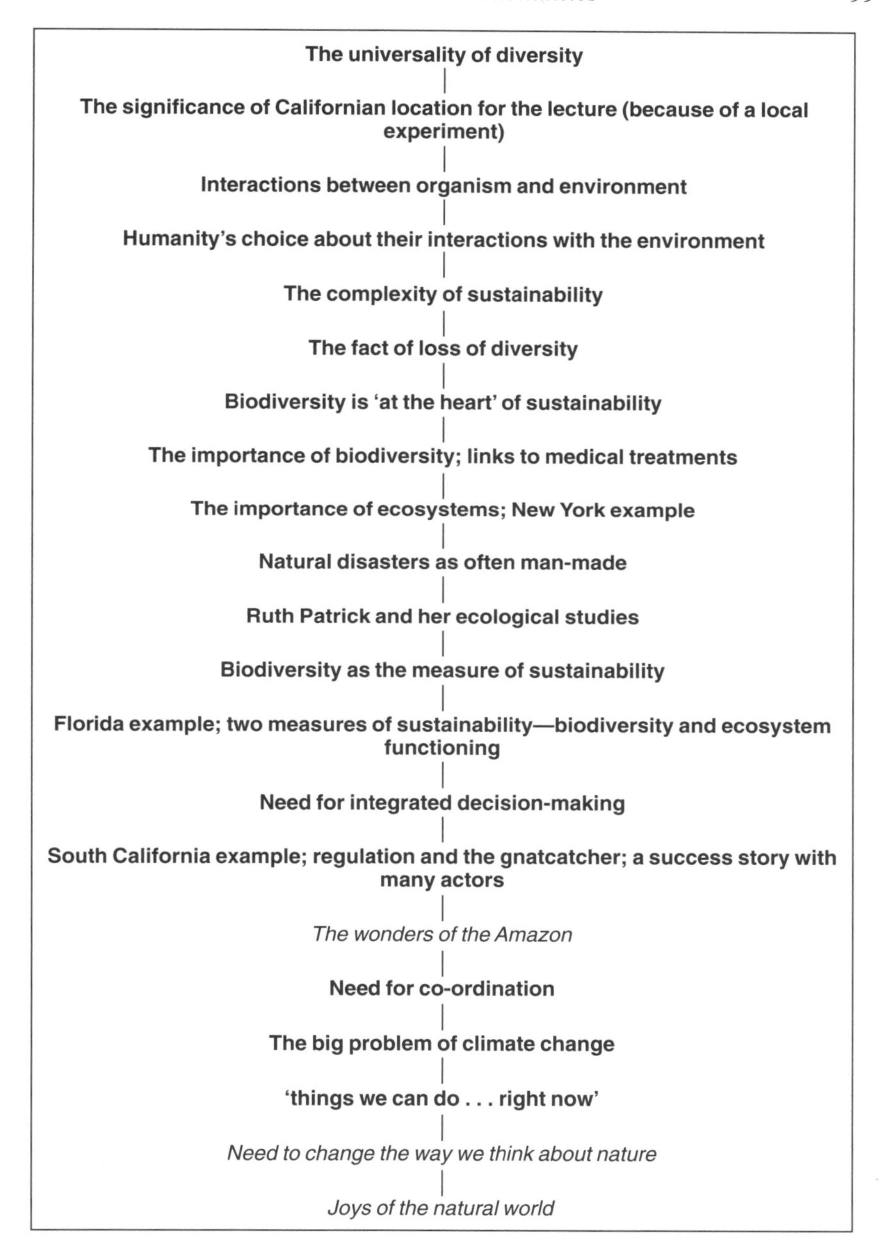

Fig. 6.1. A rhetoric line for Lovejoy's Reith Lecture 2000

Note: **Bold** indicates the discourse of scientific rationality; *italics* indicates the romantic discource.

Source: www.bbc.co.uk/events/reith_2000 accessed 30 May 2000.

	Rational scientific discourse
Metaphors	Biodiversity as library (limited use of metaphors)
Synecdoche	Single species for whole ecosystem Biodiversity for SD
Metonymy	Facts for biodiversity loss
Ethos	Professionally recognized and trained scientist
Form of closure	Scientists have the good news about what to do

Fig. 6.2. Rhetorical tropes in Lovejoy's Reith Lecture 2000

portrayed as essential in supporting the empirical study at the basis of knowledge. Without such training, environmental problems can remain unknown; Lovejoy contrasts his knowledge of Florida as a 'teenager' with that as 'Scientific Adviser', privileging the latter. In this way, scientists can act as the true voice of nature and can adopt nature's long-term perspective. The key repeated verbs that the scientific ethos uses are 'think' or 'know', and this supports the form of argumentative closure that is used: scientists know about the problem and the good news is that they also know about the solution.

While a great resource to environmentalists and a powerful discourse in society as a whole, scientific rationality is not without challenge. For the claims of scientific rationality, while widely understood within society, are not consensually accepted in all cases. As Beck (1992) has outlined and was more concretely apparent in the discussion from the floor after Lovejoy's speech, there is considerable disquiet about the claims of science and the practice of scientists (see also Hannigan 1995). These doubts currently take a number of forms. First, there is now greater social understanding of the limits to scientific knowledge and the inevitable existence of uncertainty in our understanding and management of environmental problems. Even where the claims of science as a route to the truth about these problems are accepted, there are doubts about the practice living up to the ideal. There are also doubts about the willingness of scientific institutions to acknowledge the limits to their expertise and act appropriately in the face of uncertainty. A series of environmental, food, and health crises has fuelled these concerns: BSE, nuclear power, GMOs, to name the most significant. In these, scientific institutions are seen to have been part of the policy failure, rather than the route to an optimal policy approach.

Meanwhile, within academic debates, the nature of the relationships between social and material processes has been much discussed, highlighting the social construction of scientific knowledge (as discussed in Ch. 2) and challenging the realist basis of scientific rationality. These examinations of the self-expression of scientific rationality within society at large and academic institutions are all part of the potential collapse of the modernist settlement outlined by Latour (1999). The result is that the rational scientific discourse, on

	Romantic discourse
Metaphors	Rich web of life Heart of SD Contrast with lonely wasteland
Synecdoche	Singular experience for everyone's relationship with nature
Metonymy	Amazon for biodiversity
Ethos	Prophet
Form of closure	I cannot believe that you do not agree with me

Fig. 6.3. Rhetorical tropes in romantic discourse

its own, can no longer be generally persuasive. This has led to increased use of scientific rationality alongside another basis for legitimation, particularly in the environmental context. In the case of Lovejoy's speech, he chose to combine scientific rationality with a romantic discourse.

The nature of the romantic discourse is set out in Fig. 6.3. It is a richly metaphorical discourse and full of poetic language: 'entrancing', 'shimmering', 'wondrous', 'joy', 'rich'. It is a discourse that draws on the Romantic and Transcendental tradition, which commentators have so often found within environmental texts from an Anglo-American tradition (Cantrill and Oravec 1996; Macnaghten and Urry 1998). As such it is based in the significance of the personal encounter with nature. For example, in Lovejoy's speech, he makes reference to his repeated visits to the Amazon. This particular part of the globe is frequently used to stand for biodiversity in general, a classic synecdoche within contemporary romantic discourse; Lovejoy is no exception here, emphasizing his awe and love of this place. It is also typical that the narrative moves into the first person ('I') at this point.

But to make this more than just a personal diary entry and to justify the ability of the speaker to speak for many, it is necessary to depersonalize the ethos. So the ethos adopted is that of the prophet (as again other commentators have noted in discussing this discourse; see Opie and Elliot 1996) The prophet can announce that 'our fate depends' on specific actions being taken; the prophet can demand that a change of attitude towards nature occurs. The verb that is typical here is 'believe', in contrast with the 'knowing' and 'thinking' of the scientist. Debate is reduced to questions of belief, providing a distinctive and rather unchallengeable means of closure. Scientific rationality and romantic discourse are often found together in environmentalist discourse. They seem to be able to go hand in hand with each other, provided they play complementary roles. This involves each dealing with different realms—the material and the spiritual; each involving a different mode of being—thinking and feeling; and each providing a different output—knowledge and wisdom.

Another way that the romantic and scientific discourses can be used to reinforce each other is through the development of a sense of impending crisis, a crisis that scientific rationality is called upon to resolve. The romantic discourse here plays the role of emphasizing the severity of the crisis. This almost apocalyptic romantic tone has been found to pervade many environmental texts (Killingsworth and Palmer 1992: 64–78, 1996; Hajer 1995: 80–4; Spangle and Knapp 1996). Opie and Elliot (1996) have provided a more specific analysis in terms of the jeremiad—a ritualized castigation of people for having defaulted on their bond with the Creator—which they argue is a major resource in US environmental texts. This rhetorical form can generate an emotional appeal to the audience, reinforcing the sense of a need for action. But, as Killingsworth and Palmer point out, appeals to emotion and the use of pathos are inherently risky, as they can elicit a dismissive response from the audience, rather than act as a call to action. In the absence of supporting proposals for action, this can become the soothsayer's wail. The implicit message of hope needs the response from, say, scientific rationality to be heard.

However, such complementarity is not guaranteed and romanticism can also be turned against scientific rationality. An example of this is provided by the Prince of Wales's own speech in the Reith Lecture series, summarized in Fig. 6.4. Here, the Prince of Wales relies heavily on a romantic discourse to make a very clearly sequenced and hard-hitting attack on rational science. By drawing a sharp contrast, there is an emphasis on personal experience, intuition, common sense, and traditional wisdom as alternatives to scientific knowledge. The rational scientific discourse's emphasis on empirical work and experiments is here turned into a badge of blame: science wants to treat the entire world as a 'laboratory of life' in which, presumably, we are all guinea pigs. The preferred approach outlined here is to 'restore balance' by allowing the voices of common sense and traditional wisdom into the debates. Here the scientific rationality is overwhelmed by the emotional claims of the romantic discourse. This example illustrates how scientific rationality, although still of great cultural significance, cannot be assured of the predominant position it once held.

Economic Rationality

While scientific rationality has been the subject of significant challenge within 'risk society', economic rationality remains a strong and even growing presupposition within the policy process as a whole (Dryzek 1997: ch. 6), including the environmental policy process. Indeed, at the end of the twentieth century it seemed, for a while, that economic rationality would take the dominant role in supporting environmental arguments. This represented the movement of economic rationality from other areas of public policy into the environmental domain, where scientific rationality had traditionally held sway. It was an expansion of the already extensive empire of economic rationality. For eco-

Fig. 6.4. A rhetoric line for the Prince of Wales' Reith Lecture 2000

Note: **Bold** indicates the discourse of scientific rationality; *italics* indicates the romantic discource.

Source: www.bbc.co.uk/events/reith_2000 accessed 30 May 2000.

nomic rationality has been supremely successful and is one of the bedrock 'taken-for-granted' elements of capitalist societies. Challenges to this form of rationality are often marginalized and readily subsumed.

The influence of economic rationality is one of the key themes of Foucauldian analyses, such as those by Flyvbjerg and Hajer, that seek to identify the power of the uncontested supremacy of economic interests. The key to the discursive success of economic rationality lies in congruence between a professionalized and a lay discourse on the economy. As Foucault (1984: 113) argues, this relates to: 'The manner in which economic practices, codified as precepts or recipes and ultimately as morality, have sought since the 16th century to ground themselves, rationalise themselves, and justify themselves in a theory of wealth and production.' In Fig. 6.5, these two discourses have been termed the economist's expert discourse and the discourse of corporate discretion. To illustrate these discourses at work, Figs. 6.6 and 6.7 present rhetoric lines from a statement by a politician and a book by a professional economist.

The discourse of corporate discretion refers to the language and argumentation surrounding the everyday activities of economic agents. The phrase 'corporate discretion' itself refers to Lindblom's analysis of the power that economic agents hold within the policy process as a result of their ability to undertake investment strikes or withhold the resources on which the successful implementation of public policy depends (Lindblom 1977). This institutionalist analysis of the ability of economic agents to exercise power remains a key to understanding their role within the policy process but it lacks a discursive dimension, which Fig. 6.5 can provide. This figure summarizes the way in which everyday economic discourse focuses rhetorically around the key person of the entrepreneur or businessman. This person (and it is usually a 'he') comes to be the emblem of the economy and his realm, the firm, comes to stand for the whole economy. This figure also personifies the ethos of everyday economic discourse. His everyday experience within the economy means that his viewpoint becomes a form of common sense: obviously profit-seeking is a necessary and worthwhile activity. The business even becomes the mode of argumentative closure, since 'money talks' through the medium of the entrepreneur and has the final word.

This is a coherent and self-consistent discourse but its strength lies in its uncontradictory links with the expert discourse of economists. The rhetoric

	Discourse of corporate discretion	**The economist's expert discourse**
Metaphor	Game metaphors	Newtonian dynamics; plumbing and engineering metaphors
Synecdoche	Firm stands for the economy	Market stands for the economy; economy for society; utilitarian maximization for all motivation
Metonymy	Entrepreneur is key economic symbol	Equilibrium model
Ethos	Businessman who knows profit-seeking is common sense	Expert
Closure	Money talks	We know the way the world is

Fig. 6.5. Rhetorical tropes in the economic discourses

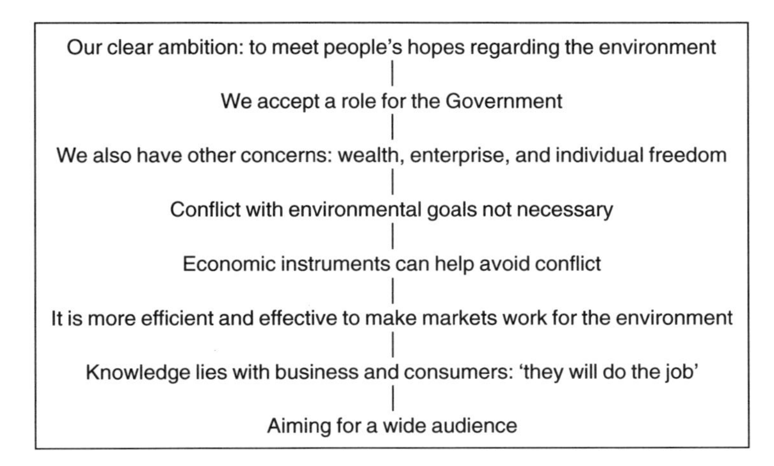

Fig. 6.6. A rhetoric line for Secretary of State for the Environment's Foreword to *Making Markets Work for the Environment* (DoE 1993)

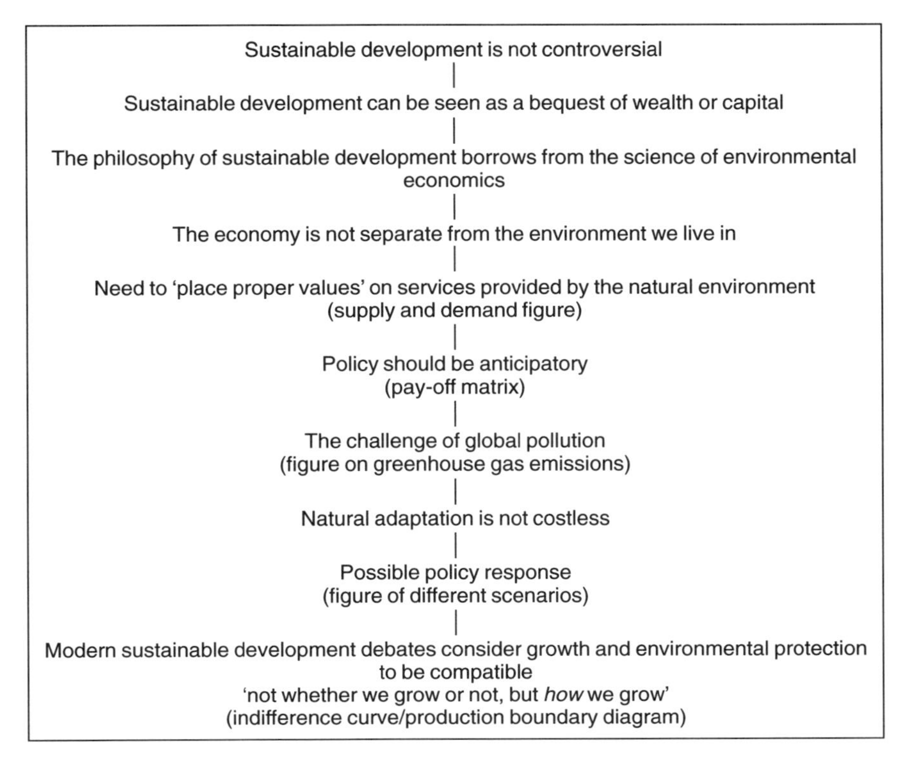

Fig. 6.7. A rhetoric line for the Introduction of *Blueprint for a Green Economy* (Pearce *et al.* 1989)

of professional economics has been fully analysed by McClosky (1994). The expert discourse is highly formalized through the use of certain models, expressed either graphically or algebraically. These models are the metonymy of the expert discourse and it becomes virtually impossible to imagine an economic text without them. They are expressed through a series of Newtonian metaphors, of flows, balance, and equilibrium. However, conceptually they are specified through the means of a set of key assumptions about human and social behaviour, commonly wrapped up in the phrase 'economic man'. These assumptions privilege utilitarian maximization above all other motivations and, in this way, also manage to privilege the market above all other institutions of the economy, the economy above all other institutions of society. The expert on the market and the economy—the economist—thus comes to stand for the key knowledge about how the world works.

The expert discourse gives authority to the trained economist, who understands the models; the everyday discourse gives authority to the experienced entrepreneur, who understands the marketplace. In either case, it is difficult to challenge the basis of the authority. This is particularly so because the two forms of authority reinforce rather than undermine each other. The one is just the expression of the other in a different context. This is apparent in the discussion about the use of economic instruments (pollution permits, environmental taxation) to achieve environmental policy goals. As the Government publication *Making Markets Work for the Environment* states: 'economic instruments harness the power of markets and make them work for the environment' (1993: p. i). There may be a mass of economic theory backing the use of instruments (and this is detailed in an annexe to the document) but these policy tools are based in the everyday operation of the market. As Secretary of State for the Environment, John Gummer (1993: Foreword) says 'they [economic agents] will do the job'.

What the expert economist does is to make visible that which is hidden within market processes, but known to the intuition of the entrepreneur. This can include the costs of environmental regulation as well as the burden of pollution on the environment. By identifying all these costs, economists are able, in that overused word, to 'balance' all costs and benefits. In order to do this, all costs and benefits are put on an equal footing and a key aspect of achieving this is that the environment and economy are treated as commensurate. The environment provides services like any other business; the environment can be treated as wealth, capital, and assets. This is, of course, a key assumption of environmental valuation techniques and their application. It also allows a highly successful formulation of sustainable development, in which environment and economy do not conflict but instead work together, which has been termed 'ecological modernization' (Hajer 1995; Dryzek 1997: ch. 8). In this formulation, economic instruments allow a context for economic activity to be set in which growth takes a less environmentally damaging path and in which environmental protection generates opportunities for profit-making. The task then becomes defined in terms of getting onto this path, a task that only business and economists can perform.

While highly successful in many arenas, economic rationality has been the subject of challenges from both environmental and social perspectives. These challenges have ranged from attacks on the conception of the environment embodied in this rationality, to criticism of the practical policy tools proposed and, further, to concern and anger over the outcomes implied by pursuing this rationality in policy situations. One aspect to these challenges is that it is difficult to combine a discourse of economic rationality with the romantic discourse that flavours so many environmental texts and pronouncements. The utilitarian language of economic discourses—both practical and theoretical—runs directly counter to the heightened language of romanticism. Since both seek to express the everyday reality of people's relationship to the environmental (as opposed to scientific rationality's more abstract conceptualization of this relationship), economic rationality is rhetorically opposed to romantic discourse. Each is seeking to present a different version of the same domain, rather than to represent different domains. Therefore, a discursive choice has to be made between economic rationality and the romantic discourse.

A classic example of this opposition is provided by the cover of the British Association of Nature Conservationists' response (BANC 1989) to the Pearce Report *Blueprint for Survival* (analysed in Fig. 6.7). The cartoon on the cover depicts bowler-hatted and pinstriped men roaming over the countryside with clipboards. They are quoted as saying: 'My heart leaps up when I cost out a rainbow in the sky' and (directly quoting the great Romantic poet, William Wordsworth), 'I gazed and gazed, but little thought at what price nature's gifts were bought'. Romanticism and economic rationality can only be rhetorically combined if the latter is recast in non-market terms. This is possible where a community-based form of economic analysis is attempted, as in some development literature and the discourse around common property regimes (Berkes 1989). It further requires that community values are seen as essentially embodying romantic values. This reformulation of economic rationality can then tend towards communicative rationality, as discussed below. Any market-led version of economic rationality is bound to be self-contradictory if cast in romantic terms.

Communicative Rationality

The term 'communicative rationality' has already been discussed in the context of Habermas's social theory. But, as noted in the discussion of collaboration and deliberation, there is also a move within policy circles to adopt and promote a practical version as a substantive policy rationality. This is based on the promotion of widespread discussion and debate within planning situations and has become widely accepted as a potential alternative to the dominant economic and scientific rationalities (Irwin 1995: 69–73; Dryzek 1997: ch. 5; Jacobs 1997*a*; Smith and Wales 2000). Note here that the emphasis is on communicative rationality as a basis for legitimation within policy circles, and not as an analytic concept within academic debate. To explore this, the legitimating role of communicative rationality within the latest UK strategy for sustainable development *A Better Quality of Life* (HMG 1999) is explored. Key quotes are provided in Fig. 6.8 and the discourse is summarized in Fig. 6.9.

In using the example of *A Better Quality of Life*, it should be made clear that this is not a document that draws only on communicative rationality; there is much emphasis on economic and scientific rationality. However, even within such a document, there are strong themes of communicative rationality. There

Metaphor	Constitutional (politico-legal)
Synecdoche	Public participation stands for the whole process
Metonymy	Forums such as citizens' juries
Ethos	True democrat
Closure	Who would deny the central important of participation?

Fig. 6.8. Rhetorical tropes in the communicative rationality discourse

Cover: photos of technology (a computer keyboard), nature (buttercups and clover) and children all set within the double-helix associated with DNA.

- Putting people at the centre
- Transparency, information, participation, and access to justice

(2 of 10 Guiding Principles and Approaches, pp. 22–3)

Public involvement is essential for a truly sustainable community. (p. 67)

People need the chance to play their part in shaping change.
(caption below a photo of people discussing in a garden, p. 68)

Sustainable development requires the participation of everyone in the UK. So it is essential that we find better ways of involving all sectors, and the public at large, in decisions. The Government has sought to adopt an inclusive approach in the preparation of this strategy, and will continue to do so. (p. 96)

The involvement of all stakeholders in a transparent reporting system will be a key tool in promoting the behavioural change that will allow progress to be made towards sustainable development. (p. 96)

The Government intends to hold a seminar . . .
(opening of final sentence of document, p. 96)

Fig. 6.9. Key quotes from *A Better Quality of Life: A Strategy for Sustainable Development for the UK* Cm. 4345 (HMG 1999)

is the emphasis on the substantial consultation process that went into the preparation of the document itself. Throughout it highlights examples of innovative practice on consultation and policy debate: the Women's Unit within the Government (now defunct); the Children's Parliament on the Environment; the Government Panel on Sustainable Development and the UK Round Table on Sustainable Development, both now replaced by the Sustainable Development Commission; and many others.

The substantial chapter on 'Building Sustainable Communities' (20 per cent of the document) identifies a range of ways of engaging with communities:

- voting, Local Agenda 21;
- consultation within land-use planning;
- the Environment Agency's Local Environmental Action Plans, which include consultation with stakeholders;
- the operation of the New Deal for Communities; and
- the emphasis on voluntary work encouraged by the Home Office's Active Communities Unit.

Even the proposed indicators have to take account of communicative rationality; two proposed indicators are 'community spirit' and 'voluntary action'. This is the presence of communicative rationality within a key policy document, but how does this discourse work?

The first point to note is that, unlike the other two rationalities discussed in this chapter, communicative rationality is not a technical discourse based in specific expertise, but a politico-legal discourse based in the philosophy of rights. Its metaphors are constitutional: deliberation, juries, citizens. This is why there is such a fundamental tension between this rationality and the others, and why, when put together, communicative rationality so often undermines the others (as is discussed in Ch. 10). It also makes this rationality a difficult one to avoid in democratic societies. It connects up to a much broader discourse about the state and society, and the nature of democracy. As such it is a social 'taken-for-granted', part of 'motherhood and apple pie' and difficult to oppose at the level of statement. This remains the underlying ethos of communicative rationality and informs all the other rhetorical tropes: participation stands for the whole policy process, the benefits of participation become the undeniable form of closure, and participation forums become the metonymy of communicative rationality.

But the communicative rationality discourse is not just about constitutional rights; it draws on that broader discourse in order to develop a distinctive policy approach. In this it has common ground with the other rationalities; they are all practical policy rationalities. As such, communicative rationality seeks to present an optimistic picture of policy practice and how it may develop new modes of engagement with communities. The use of individual examples of success stories is widespread, usually allied with an expressed hope for such practices to spread and become the norm, not the exceptional example.

But, like many optimistic but radical outlooks, communicative rationality is not firmly grounded in methods for achieving this (as Chs. 4 and 5 have discussed). This leaves communicative rationality reliant on the organic development and spread of appropriate techniques, models, and approaches for encouraging deliberation and debate.

In the meantime, the emphasis is on providing opportunities for communication and heralding each case of consultation positively. Talk becomes a goal in and of itself. As has been explored above, one problem with this is identifying the interests, particularly professional interests, who would benefit from the promotion of communicative arenas. The links to a broader democratic discourse suggesting that it is in all our interests to engage in such communication. But while democracy may be in all our interests—a genuine common interest—it is not so easy to make the transition to arguing that communicative rationality within the policy process is in all our interests. The thorny problems of collective action, conflicts of interest, and the interests of the bureaucracy all militate against this. There are similar problems in combining the romantic discourse with communicative rationality, except in situations where the community or group of stakeholders holds fundamental environmental values. This heightened mode of expression may find a resonance with some stakeholders, but this cannot be guaranteed and, indeed, may actually be antagonistic to others' preferred way of expressing themselves. In these situations, it could be that a romantic discourse would actually undermine the pursuit of communicative rationality in practice.

Comparing and Contrasting Substantive Rationalities

These three rationalities can all be found justifying specific policy approaches. Each seeks to shape the policy agenda in a particular direction, and each provides its own specification of what constitutes environmental problems and preferred or, at least, acceptable policy solutions. These are summarized in Fig. 6.10.

As can be seen, each rationality constructs the environmental policy agenda quite differently. For scientific rationality, environmental problems are to be found in physical, material reality and it is the role of science to understand them, providing knowledge that can then lead policy approaches. The key question becomes: how much do we know about a particular environmental issue? What are the uncertainties and how can research reduce them? Is this knowledge being properly translated and used in the policy arena? Such questions have shaped much of the discussion around recent public health and food issues such as GMOs, BSE, mobile phones, and foot-and-mouth disease. While the precautionary principle is often presented as providing a basis for policy moving ahead of confirmation by scientific knowledge, actually it continues to

	Scientific rationality	**Economic rationality**	**Communicative rationality**
View of the environment	Physical reality Object of scientific inquiry	Resource Object of consumption Context for economic processes	Socially constructed Interface of the physical and social Quality of life
Nature of environmental problems	Arising from lack of understanding and knowledge; leading to poor management	Arising from unpriced, overused resources, and lack of property rights; not incorporated in economic decision-making	Arising from inadequate stakeholder involvement, rejection of lay knowledge, and insufficient environmental education
Preferred environmental solutions	Based on sound science Knowledge-led	Market-based instruments Introducing property rights and quasi-market pricing	Consultation with stakeholders Visioning, etc. Consensus building

Fig. 6.10. Comparing and contrasting the three rationalities

be based on this rationality. Certainly it does represent a critique of contemporary scientific practice (O'Riordan and Cameron 1994) but the precautionary principle does not reject scientific rationality. Rather it buys time for scientific expertise to develop its knowledge base. It prevents the status quo changing so irreversibly that policy based on 'sound science' would come too late. Yet it looks to scientific rationality to provide the first indications of where cause–effect relationships may be leading to environmental problems.

Economic rationality, however, focuses on the inevitable dynamics of social, particularly economic processes. It therefore seeks to view the environment through this lens and, in policy terms, to incorporate the environment within the ongoing stream of economic decision-making. It sees the environment as both context for and an element within that decision-making, but both these dimensions are largely invisible or silent because of the lack of a pricing mechanism or existing property rights. From this, the agenda logically follows in terms of finding ways for pricing or valuing the environment and creating rights to the use of and benefit from environmental 'resources', 'assets', and 'services'. The environment becomes environmental capital, alongside other forms of capital. The issue of biodiversity becomes the value of the genetic pool, its protection through assigning rights, and investment in its future through economic exploitation of some sort. This may take the form of multinational pharmaceutical companies patenting specific DNA sequences and developing commercial applications therefrom—bioprospecting or biopiracy as it is variously termed—or local communities declaring common ownership of species and habitats as a basis for locally based extractive industries. In either case, though, this is an interpretation of economic rationality at work, albeit filtered through particular ideological colours.

Communicative rationality, as intimated above, begins from a different standpoint, one concerned with social and political processes, with citizens' rights and duties, and processes of consultation and empowerment. It sees the environment and environmental problems from this standpoint, so that the environmental and social are always seen as intertwined. Environmental attributes, including problems, are socially constructed and intimately involved in people's quality of life, in some cases in their livelihood. Involvement of actors, groups, and communities is, therefore, central to any discussion of the environment and any potential policy approach. Environmental risks, such as GMOs or telecommunication masts, can only be seen in terms of engaging affected communities or 'stakeholders' in discussions, with a view to producing informed acceptance or rejection of that risk. Scientific knowledge (or its lack) is only an input into such processes of engagement, and economic interests are only one stakeholder among many. As might be expected, this communicative rationality has proved extremely challenging to the other two rationalities discussed here.

Combining Substantive and Procedural Rationalities

Before exploring how the three rationalities operate in specific case studies, it is worthwhile considering how these substantive rationalities relate to the procedural rationality of the policy process (discussed in the previous chapter). Such a discussion can help our understanding of why some substantive rationalities will find a place more readily within the discourses of environmental planning.

Scientific rationality, out of all three, meshes most readily with the rational policy process discourse. It supports the idea of the policy process as neutral and disinterested, drawing on outside expertise for specialist advice. The neutrality of scientific rationality as an objective path to knowledge supports the rational policy process discourse in its claims to objectivity. Furthermore, the use of scientific rationality as an external input fits well with the idea of the policy process as broken up into stages, which was seen to be an integral element of procedural rationality. The way that the two rationalities mirror each other suggests that procedural rationality is the policy-world equivalent of scientific rationality. Indeed some national cultures speak of a scientific policy approach, when they mean strict adherence to procedural rationality. They are both, in their different contexts, seen as the epitome of rational logic, the one in relation to understanding the material world, the other in relation to managing the tasks of government policy. Finally, these two rationalities are complementary because they do not compete for responsibility of the same domain. Scientific rationality only acts as an input into procedural rationality and does not challenge for control of policy decisions. For this reason, it is in the interests of policy officials to refer to scientific rationality. Control of decisions remains in their hands but problems in outcomes can be referred back to problems of the advice that has been received. The case study of air quality management in the next chapter develops this theme.

Economic rationality, by contrast, has a less clear-cut relation to procedural rationality. In so far as economic rationality can be turned into a form of procedural rationality, there is a symbiosis. This happens, for example, with environmental valuation techniques and evaluation methods such as cost-benefit analysis. Here the language and reasoning of economic rationality, particularly the expert version, lends its support to the claims of the policy process itself to be rational. The same logic of stages in decision-making and the same self-presentation as neutral apply in both cases and reinforce each other. However, there is also the potential for a struggle over control of decision-making. In so far as economic rationality—in both its expert and corporate discretion versions—suggests that decision-making should be left to economic actors, this can prove problematic for policy-makers. In some circumstances, this can prove a welcome way to shift responsibility for outcomes but it can also mean that the remit of control for policy-makers is too constrained. The case study of housing land policy in Ch. 8 illustrates this point. Economic rationality becomes an alternative to procedural rationality, rather than an input. Even where economic metaphors and language are used within the policy process, the claims of economic rationality can become a more substantive engagement with issues of control and responsibility, in turn raising issues about regulation versus the market. This was shown to be the case with environmental planning gain, in the analysis by Whatmore and Boucher (1993).

Finally, there is the engagement of communicative rationality and procedural rationality. It can be argued that the need for the outcomes of procedural rationality to achieve public acceptance requires the policy officials to engage with communicative rationality (Murdoch, Abrams, and Marsden 1999, 2000). However, consideration of the discursive claims of both rationalities shows the potential for severe conflict, for they both centre around claims as to who should be in control of decision-making. Procedural rationality provides a basis for legitimating the claims of policy officials, while communicative rationality has at its centre the claim that decision-making can only be legitimated with regard to broad-based participation. Communicative rationality, with its constitutional language, raises the issue of where political authority should lie. Connecting this with procedural rationality then challenges the self-presentation of neutrality within the rational policy process. In effect, it politicizes the policy process (Grant 1994). This provides a further reason why policy-makers may, at times, be reluctant fully to engage with a discourse of communicative rationality. Healey has highlighted how this happened in the past. Discussing the complexities of stories arising from more inclusionary modes of planning, she states, 'In the rational planning mode, these were ignored' (1996: 28). This again suggests a reason why changing environmental planning practice in line with communicative rationality may prove problematic, reinforcing the discussion of Chs. 4 and 5. Following Hillier (1997: 32), planners may seek to subsume communicative rationality within procedural rationality: 'Planners, therefore, often seek to enrol other actors into their representations. Their goal is mobilization; acceptance of the planners' representations as legitimate by local residents.'

One counter-argument (already alluded to in Ch. 5) is to point to the potential emergence of a new strand of planning profession, whose practice is legitimated by communicative rationality and, procedurally, based in expanding the opportunities for participation (Evans and Rydin 1997: 64–6). Here planners can use communicative rationality to help create a self-identity. The fact that this discourse is used across the boundary between the state and civil society is itself a reflection of the professionals' desire to build bridges with communities and to distance themselves somewhat from conventional forms of policy talk, based in procedural rationality. The case study of Local Agenda 21 in Ch. 9 will consider these points further.

Conclusion

These three substantive discourses of scientific, economic, and communicative rationality are available to actors to use in discursive strategies in a variety of policy relevant arenas. While there is considerable latitude open to actors to be creative in these discursive strategies, the overall patterning of these discourses does present constraints as well as opportunities. The patterning delimits the domain of the acceptable, or at least the readily and widely acceptable. It also specifies linkages between forms of argumentation and the likelihood of swaying a particular audience: that is, an environmental argument couched in terms of economic rationality is more likely to persuade an audience of business stakeholders. Furthermore, the three discourses vary in terms of how they can combine with the procedural rationality of the policy process itself. How these rationalities work in practice will be explored through three policy case studies in the next three chapters.

7

Air Pollution Control and Air Quality Management

Among environmental issues, pollution is one that has been subject to considerable 'unpacking' in terms of its social construction. Mary Douglas's (1966) work on the construction of matter 'out of place' as 'pollution' and the development of social institutions of taboo and control on that basis is rightly regarded as seminal within anthropology. Her own and subsequent work on the construction of risk has developed this approach further, particularly within the anthropological framework of cultural theory (Douglas and Wildavsky 1983; Adams 1995). However, there has been little work that actively links this social constructivist perspective to an analysis of contemporary planning practice. Adams's work takes policy as its object of study, deconstructing and classifying it according to a cultural theory typology, but he does not explore the details of its practice. And yet pollution control provides an ideal case study for linking the construction of the policy problem with institutional arrangements for handling that problem. This chapter therefore takes pollution control, and in particular air pollution control, as its focus. It considers the policy framework in Britain and its links with European policy, and ranges across industry-specific and locality-wide policy practice.

Air Pollution and Scientific Rationality

Contemporary policy concerning air pollution is a classic case of reliance on scientific rationality. Pollution is defined as such by reference to the harm it causes:

> 'pollution' shall mean the direct or indirect introduction as a result of human activity, of substances, vibrations, heat or noise into the air, water or land, which may be harmful to human health or the quality of the environment, result in damage to material property, or impair or interfere with amenities and other legitimate uses of the environment. (Article 2.2 of EU Directive 96/61 on Integrated Pollution Prevention and Control)

The extent or even existence of such harm, damage, or impairment is, in most contemporary circumstances, only revealed through scientific expertise in measurement and diagnosis. Reliance on the senses of everyday experience is insufficient to reveal levels of pollution or air quality. Visible emissions may be harmless to humans and the environment. Invisible emissions may be toxic or represent a growing environmental problem. The Greater London Authority's draft London Air Quality Strategy (2001: 3) argues that, while London is the most polluted city in the UK, 'today, most air pollution is invisible'. Scientific expertise, therefore, seems to be a necessary element within the air pollution policy arena. This expertise takes a number of different forms.

First, there is the role of scientific expertise in revealing the extent of pollution. Science is here seen as revealing the environment, the natural world, to the social world. It does so in a very specific way. The categories used refer to the extent of the presence of specific chemical compounds in the atmosphere. These compounds provide the primary classification system used and, as such, require interpretation for those without a chemistry training. This is even more the case since pollution usually involves a cocktail of such compounds. Currently European policy on controlling industrial emissions to air recognizes thirteen sets of chemicals; twelve of these are 'families' of chemical compounds while one covers all substances and preparations with proven carcinogenic or mutagenic properties, or likely to affect reproduction (Annexe 3 of EU Directive 96/61/EC). British air quality monitoring collects data on seven compounds, which are considered the relevant pollutants.

The measurement of these compounds carries with it a specific attitude to space and time. The monitoring of pollution can take place only at specific sites and this is the case whether emissions themselves are recorded or their presence in the atmosphere beyond the emission point is recorded. The actual measurements taken are usually very sensitive to the precise location of monitoring equipment so that the scientific discourse on air pollution takes on a very specific form: so much of a particular chemical compound at a specific spot. Furthermore, this measurement can be presented in a number of different temporal ways. Real-time recording is possible providing a continuous record of the presence of pollutants. But such technology is expensive and the quantity of data generated cannot always be justified in cost terms; cheaper approaches with periodic measurement may be adopted. And when the data is presented, a summary over time is usually required (although there are exceptions, as in Vienna, where a real-time public display indicates the emissions from a waste incineration plant). Summary data has to decide whether to present the data for instances in time or over periods of time (of hours, days, etc.) and how frequently within a longer period of time (a two-hour period each day, six periods per hour, etc.). For example, sulphur dioxide in the atmosphere is measured in terms of the average for 15-minute periods; nitrogen dioxide as an average over hourly periods; carbon monoxide is measured as an 8-hour running average; and particles (PM_{10}) as a 24-hour running average.

Clearly measurement per se has a limited purpose, essentially a descriptive purpose. This is not a value-free purpose, since the very selection of compounds for measurement as pollutants implies a normative perspective. But this is enhanced by the 'judgement' of these measurements against standards. Currently British environmental policy uses a number of standards derived from the World Health Organisation, the European Union and its own Department of Environment, Food, and Rural Affairs. However, because of the force of European law, the EU standards are coming to dominate. DEFRA has increasingly modified its standards in line with the EU norm. In doing so it has not only had regard to European law (a form of procedural rationality) but also to the views of its own Expert Panel who have provided the access to the scientific language in which both European and British norms are expressed. These norms are, in turn, supported by medical epidemiological expertise about the links between the level of a particular pollutant and the impact on human health. Thus the government leaflet on *Air Pollution—What it Means for Your Health* justifies the normative description of air quality in terms of its bands of "low", "moderate", "high", and "very high" thus: 'The breakpoint between the "low" and "moderate" bands uses the health based air quality standard . . . based on the advice of the UK's Expert Panel on Air Quality Standards . . . The breakpoints between the three higher bands—"moderate", "high" and "very high"—are recommended by the Department of Health's Committee on the Medical Effects of Air Pollutants' (DEFRA 2001: p. x).

This does not fully describe the remit of scientific rationality and the role of scientific expertise in air pollution policy. For as well as constructing the measurement of pollution and enabling its connection to normative judgements on the acceptable level of pollution, scientific rationality is drawn upon to provide a dynamic picture of how pollution works. The emphasis here is on presenting a model of how pollutants move within the atmosphere and are concentrated or dissipated. This in turn involves knowledge claims about the dynamics of the atmosphere, including air currents and other aspects of meteorology (or, in lay terms, weather). Modelling the interconnections between emissions, changing weather, and the nature of the atmosphere is presented as providing the knowledge base for pollution policy. This has three implications. First, such knowledge should enable a better understanding of how a specific emission is dissipated or becomes concentrated and, therefore, how it should be judged in terms of normative standards for pollution. Second, it holds out the prospects for a system of forecasting, whereby changing weather can be translated into expected changes in pollution. This can support advice to the general public and health professionals about short-term pollution impacts. Third, it can provide an analysis of whether current policy standards are likely to be exceeded over the medium term and, therefore, whether additional policy action is necessary to prevent breaches of these standards.

These uses underpin the relation of scientific knowledge about pollution to the regulation of emissions and to the provision of information to the public. These will be dealt with in turn. The regulation of pollution—through pollution control—invokes procedural in addition to scientific rationality. These are connected quite directly in that scientific categories justify particular modes of regulation. Scientific rationality distinguishes between point-source and diffuse pollution. Point-source pollution originates in specific individual sites, such as an industrial complex; diffuse pollution arises from many sources each of which may be relatively minor, such as the many individual uses of motor vehicles or, in the past, many individual coal fires in homes. According to a realist perspective, these different types of pollution have such different physical natures that they require different regimes of regulation. The construction of the pollution problem justifies the organization of state involvement to deal with that problem. This, in turn, has led to two quite different sets of institutional arrangements.

Procedural Rationality and Pollution Control

Taking the regulation of point-source pollution first, regulation takes the form of authorizing individual industrial plants to emit a specified amount of certain compounds. Pollution is defined as exceeding these authorized limits; pollution is not emissions per se but emissions beyond these limits. Scientific authority is, therefore, centrally involved in defining not only the compounds involved but also the permissible levels. These are known as the 'emission limit values'; they are specified in terms of a certain maximum mass of each of the compounds per period of time. The authorizing body varies with the nature of the process subject to regulation. Currently, local government, subject to guidance by the Environment Agency, regulates simple processes that emit only specified substances into the air (Regime B) and also processes with a lesser potential to pollute, even where that involves emissions to air, land, and water (Regime A2). More complex processes, involving a range of potentially more significant environmental impacts, are regulated by the Environment Agency (Regime A1).

While scientific rationality is central to defining pollution, the process of controlling pollution takes its legitimacy from the way in which decisions are taken by regulators. A central concept here is that of the 'Best Practicable Environmental Option'. BPEO refers to the search for a combination of emissions to land, air, and water, which in combination produce the least environmental impact. It was adopted as the key guiding principle in British pollution control following the Fifth Report of the Royal Commission on Environmental Pollution. Clearly scientific authority is involved here in identifying the environmental impact of different combinations of emissions. But procedural rationality is also being invoked, in the idea of a more holistic approach to pollution and the definition of a set of procedures to achieve that holistic goal. 'Integrated Pollution Control' (IPC) was coined as a description of this particular policy approach. Integration is a term deriving its authority from procedural, not scientific rationality. It emphasizes the contribution of a policy approach that can comprehensively survey and synthesize. These are the hallmarks of the rational policy approach and represent an extension of the simple pollution-control approach, which had regulated on the basis of scientific knowledge of individual pollutants.

More recently, it has been recognized that the IPC approach itself is limited by its emphasis on handling the outputs of the production process, considering these different outputs and manipulating relative quantities of each to the benefit of the environment. This limits the scope of procedural rationality to 'end-of-pipe' solutions. For a rationality based on notions of comprehensiveness and an ability to take on board all aspects of a problem, this is frustrating. Therefore, pollution control has increasingly looked inside the production process also, considering how changes to those processes may reduce emissions in the first place. This has been described as favouring 'upstream' measures, rather than 'end-of-pipe' control (European Climate Change Programme EU COM (2001) 580 final: 7). This rather more attractive metaphor suggests pollution policy as moving closer to the source of pollution problems and sees production (and its associated pollution emissions) as a dynamic, naturalized stream, which presumably can be diverted, dammed, or otherwise influenced. The scope for pollution control activity seems much larger than that available at the end of a pipe.

The bigger picture is now formally recognized with the shift from IPC to IPPC, Integrated Pollution Prevention and Control. EU Directive 96/61/EC provides the framework for this regulatory approach, which is being fully implemented in Britain in the period up to 2007. This has involved extending not only the scope of regulatory activity from control to prevention but also extending the numbers of sites of regulatory activity, from some 2,000 processes to around 6,000 installations in England and Wales. This is a considerable increase in regulation, both numerically and in terms of focus.

However, even before the shift in formal regulatory framework, the working practices of pollution control were beginning to take on more than authorization of end-of-pipe emissions. For, while the regulatory framework emphasizes the decisions taken by the regulator, the authorization of emissions, and the potential for enforcement of breaches, daily working practice is rather different. Pollution-control practice involves regulators in frequent contact with those being regulated, usually at the site of the production process. The setting of authorizations cannot be done without information from those managing the production processes, both at the level of the individual plant or site and at the more general level of the guidelines issued by the Environment Agency. Indeed it is relevant that these guidelines are issued under the headings of different industrial processes, a shift from scientific to industrial categories as the main source for classification.

Given these close contacts between regulator and regulated, authorization processes typically involve significant negotiation around the themes of what could be done from the industrial perspective as well as what should be done from an environmental perspective. Such negotiations begin from the existing organization and technology of production processes but have increasingly addressed the question of whether that organization and technology could be altered to reduce the level of emissions and, hence, the problematic facing pollution control. This is what has now been given formal recognition through IPPC.

Of central significance for both IPC and IPPC is the attitude to technology. This is incorporated through the key concept of Best Available Technology, highlighting the commitment to achieving procedural rationality through reference to the cutting edge of relevant technological knowledge. The comprehensive knowledge base of rational planning, therefore, has to extend to technology also. However, this is not an unconstrained commitment to best technology. Before the introduction of IPPC, the concept was referred to as BATNEEC, 'Best Available Technology Not Entailing Excessive Cost'. This encapsulated the intention to implement state-of-the-art technology to prevent pollution but only after having regard to the economic consequences. Pollution control, therefore, had to take note of economic rationality also in justifying its decisions. Since IPPC, this economic element has been internalized within the definition of 'available': ' "available" techniques shall mean those developed on a scale which allows implementation in the relevant industrial sector, under *economically* and technically viable conditions, taking into consideration the *costs* and advantages' (EU Directive 96/61/EC Article 2.11, my stress). Here the combination of scientific and procedural rationality that defines pollution control has to have regard to economic rationality also.

Economic Rationality and Pollution Control

Pollution control is sometimes considered as a case of conflict between the state regulator and the regulated business (Gouldson and Murphy 1998). Conflict arises because industry seeks to avoid or minimize the impact of regulation due to the economic costs involved, while pollution control officers see their role in terms of achieving regulation in pursuit of scientifically justified standards. As such pollution control might seem to be a case of conflict between a combined scientific and procedural rationality—represented by the pollution control officer—and economic rationality—represented by the industry. But all studies of the minutiae of industrial pollution control tell of a close relationship between experts from both the industrial and governmental sectors, over the application of technology and its pollution implications. Reference has already been made to the way in which pollution-control practice involves a degree of negotiation between regulator and regulated. Pressures for co-operation arise because industry needs authorization to emit in order to undertake its economic activities, and the regulators need the technical expertise within the industry in order to determine the appropriate level of emissions.

Therefore the resulting discourse has to include significant elements of economic rationality, as in the case of the BAT concept, referred to above.

The question then becomes one of how economic, scientific, and procedural rationality are combined in the discourse of pollution control. This cannot simply be a matter of balance, a bit of one and more of another. Discourses have to find a point of compatibility to be combined, a way in which the justification provided by one discourse of rationality can be reinforced rather than undermined by the other. In the case of pollution control, the key factor is that the discourse draws on claims from a variety of sources to practical rather than theoretical expertise. This is the case with both the economic and scientific dimensions: the economic rationality relates to understanding the basis of profitability of the firm, the scientific rationality to the application of technology to the particular production process. So combining these rationalities involves combining claims that both arise from a practical engagement with the activities of the firm.

The result is a reworking of both rationalities. Scientific rationality becomes applied, practical, and specific to the particular plant and site—technological rationality. Economic rationality becomes defined with reference to the application of this technology and, in that process, becomes reworked from even the practical economic discourse of corporate discretion. It needs to engage much more specifically with the impact of technology on economic processes. The technology needs to be viable to be installed but the key terms of discussion relate to its effectiveness, whether in tackling pollution at the 'end of the pipe' (that is, at outlets of emissions) or considering how the production process itself may be modified to reduce pollution 'upstream', at source. In this way, technological expertise penetrates the economic discourse. This mutual reworking of economic and scientific rationalities is evident in the redefinition of BATNEEC as BAT and the guidance provided by central government for determining BAT, which involves both environmental and economic assessment on equal terms (DETR 2000: ch. 9).

It is not necessarily easy to work out from this whether the resulting technological changes will be driven more by environmental or economic concerns; both are integrated into the key concepts. Much will depend on the specific norms that develop through IPPC practice and the way this affects relationships between regulator and regulated. In the past there has been considerable unease about the possibility of regulatory capture, in which the regulator takes on the norms of the businesses and control becomes, as a result, lax. There are clearly attempts being made at the European Union level to signify a preference for norms that favour stricter environmental protection. In particular, there have been two attempts to use economic rationality in the pursuit of environmental goals.

The first of these attempts concerns the renewed emphasis on the Polluter Pays Principle, restated in Section 1 of EU Directive 96/61/EC. This principle uses the argumentation of neoclassical environmental economics (discussed in Ch. 6) to identify environmental impacts as externalities of economic activities (here production activities) and therefore essentially as costs that are spread among the wider public rather than paid for by the firm (or consumer) generating them. The implication is that the firm (and eventually the consumer) should pay. Therefore pollution prevention and control can be expected to impose a cost on the polluting installation; the interpretation of BAT will dictate how far that cost is limited by profitability concerns. The second attempt takes the Polluter Pays Principle one stage further by introducing the concept of 'environmental liability'. In announcing the Commission's intention to prepare a directive on this topic, Commissioner Margot Wallström stated, 'the time has come for the EU to put the polluter pays principle into practice' (EC Press Release IP/02/127, 23.1.02). Under such a liability, those operating risky or potentially risky activities would be financially responsible for the environmental damage caused or restoration required. While principally focusing on damage in terms of biodiversity, watercourses, and land contamination, such a liability again seeks to shift the balance between environmental protection and economic profitability in favour of the former. Again, it uses economic rationality to internalize environmental concerns by making the polluter bear the full costs of environmental damage. If it succeeded then pollution-control practice would be considerably altered; businesses would be operating with norms already influenced by a consideration of the costs of environmental damage even before negotiations on authorizations commenced. Environmental protection would be institutionalized within economic institutions.

Procedural Rationality and Air Quality Management

Pollution control finds itself in a quite different situation with regard to diffuse pollution. Here the structure of the problem can amount to a 'wicked' problem for policy-makers, in the sense that it approximates quite closely to the classic prisoner's dilemma scenario. The structure of the problem largely arises from the generation of pollution through multiple individual uses of cars and other road vehicles. Dealing with it is a collective action problem at three levels:

- at the level of the generation of diffuse pollution, where there is no individual benefit in anyone taking unilateral action to reduce emissions;
- at the level of NGO lobbying action on the problem, where it will be more difficult to generate political activism on diffuse air pollution than on other related issues; and
- at the level of state action itself, where the electoral costs of taking action, which will adversely affect a specific and clearly identified group (such as car drivers), outweigh the generalized benefit to many of improved air quality, a benefit they may not even notice and is unlikely to be politically salient.

Because of this structure to the problem, co-operative solutions are unlikely to arise spontaneously and because of the electoral cost of imposing a solution, public-sector action to deal unilaterally with the problem is also likely to be delayed. It requires an institution with a very particular set of incentives to take action on this problem.

From a policy-maker's perspective, this means that it is often necessary to find an interpretation of the problem and the appropriate response that will justify the limited scope of action by the bureaucracy. The discussion around such diffuse pollution has, therefore, been reframed in terms of management rather than control. The expectations involved in such a reframing are clearly downgraded. In addition, reframing moves the policy action from the technological arena discussed above, into the more general world of policy management and strategic action. This can be seen at work in the government's recent Air Quality Strategy, which is strongly couched in terms of procedural rationality. Figure 7.1 sets out the rhetoric line for both the 1995 draft and the updated 2000 version. This emphasis on procedural rationality can be seen as a means of shifting attention away from the substantive problem itself towards the way in which the problem is managed. However, the increasing importance of the substantive scientific rationality over time, from 1995 to 2001, suggests that this may be a time-limited strategy.

Nevertheless, procedural rationality will always remain significant. The National Audit Office report on the development of the Air Quality Strategy emphasized the way in which it met the Cabinet Office criteria of 'Professional Policy Making for the Twenty-First Century' by following the appropriate procedural guidelines (NAO 2001). These required a policy-making process to be:

- Forward-looking: taking a long-term view
- Outward-looking: taking account of factors in the European and international situation
- Innovative and creative: open to the comments and suggestions of others
- Using evidence: using best available evidence from a wide range of sources
- Inclusive: taking account of the impact of the policy on different groups
- Joined up: looking beyond institutional boundaries
- Evaluative: building systematic evaluation into the process
- Reviewing: keeping established policy under review, and
- Learning lessons: learning from what works and what does not.

The NAO (ibid. 2–6) report found that the British Air Quality Strategy 'acted to obtain the best evidence available at the time on the effect of air quality on health', 'conducted an evidence-based assessment of the options', and 'established arrangements to implement the Strategy and monitor progress'. This was framed as a favourable assessment of air quality policy-making.

Given the original construction of the policy problem of air pollution, it is not possible to be completely distanced from scientific rationality, but the role of this rationality alters. Research into air quality management in London reveals how scientific rationality interacts with procedural rationality in the context of diffuse urban air pollution (Rydin 1998*b*). The central bureaucrats

Air Quality: Meeting the Challenge
Department of the Environment 1995

Introduction
Starts from UK Sustainable Development Strategy and argues 'that further action should be taken now'
|
The Strategic Framework for Clean Air Policy
An emphasis on the role of standards, setting them, and their scope
|
National Standards and Targets
Pollutant-by-pollutant account of 'progress so far'
|
Local Air Quality Management: air quality management areas
Policy 'firmly based' on air quality rather than emissions
'Most appropriate that primary responsibility . . . should rest with local authorities'
Requirement for local air quality appraisal
Enabling powers for local air quality management areas
Resulting in integrated air quality management
|
Air Pollution and Transport: an action programme
List of past achievements
Statement of need for a wider strategy
Links between land use planning and local transport strategies
Shift to discussion of:

- 'obligations' of fleet operators*
- **'technology, emissions, fuels, standards'**
- 'enforcement'
- 'voluntary action and guidance'*

|
Monitoring and Public Information*
The objectives of monitoring programmes
The link to public information*, particularly during **pollution episodes**

Fig. 7.1. A rhetoric line for British Air Quality Strategies 1995 and 2000

a
Note: Underline indicates the rational policy process discourse; **bold** indicates the rational scientific discourse; *indicates a reliance on actors outside the state.

involved, the environmental health officers, supported an interpretation framing the problem as one of air quality 'management' rather than the more assertive pollution 'control'. They also argued for a central role to be given to monitoring and, subsequently, to modelling the problem, with more direct policy action only occurring subsequently. This framing had several advantages. It advanced the interests of the environmental health officers by placing the interim 'solution' of monitoring firmly within their own expertise. It also justified delaying more interventionist action on pollution, since monitoring and even modelling had to come first; this indeed is the line now taken by the British Air Quality Strategy. Furthermore, it fitted with the techno-scientific ethos that is prevalent in pollution-control discourses (see above) and it also offered a key resource to local authorities, that of information. In this way the emphasis on procedural rationality—through strategic management—and an

Working Together for Cleaner Air*
Department of the Environment, Transport, and the Regions 2000

Introduction
Examines the purpose and audience of the strategy, and establishes its 'guiding principles' and scope
|
Setting the Scene
Links to the Sustainable Development Strategy and other government initiatives
Summarizes developments in air quality
|
Legislation and Policy Framework
National and International policy context
|
Air Quality Standards and Objectives
The setting of standards and objectives
Details for individual pollutants
|
Delivering Cleaner Air
Details the role of government, industry, agencies, local authorities, business and individuals*
Establishes links to transport
|
Next Steps
Future work on **health**, *cost/benefit analysis*, **monitoring**, **PM_{10}, NO_2, new pollutants**
|
Technical Annexe

Fig. 7.1. (*cont.*)

b

Note: Underline indicates the rational policy process discourse; **bold** indicates the discourse of scientific rationality; *italics* indicates the economic rationality discourse; *indicates a reliance on actors outside the state.

attenuated form of scientific rationality—through monitoring and modelling—can be seen as a way of avoiding the 'wicked' problem of diffuse pollution.

So how can action progress on diffuse pollution? As indicated above, it requires a particular set of incentives for policy action. This can be explored by considering the issue of the introduction of road pricing or congestion charging, as will happen in London from 2003. In the days before the Greater London Authority (GLA), it was difficult to imagine such a policy arising from local government action. There were considerable collective action problems facing the various local authorities (the boroughs), which prevented them coming together to instigate such a scheme. For the introduction of road-pricing, while perhaps reducing air pollution in a borough, would do so only at the cost of reduced traffic and the associated economic benefits. In addition, while residents might benefit from the improved environmental quality, they would also be affected by the restrictions on traffic, along with other key local political actors—local businesses. Economic rationality would seem to act against such a scheme, both in terms of the impact on local economic development and the costs of introducing it. There was no existing procedural rationality for the scheme to connect to; instead this would be a policy innovation. Scientific rationality might support action for improving air quality but could offer no very specific justification for this policy option. And communicative rationality was more likely to inhibit rather than encourage such a scheme, given the distribution of costs and benefits.

This structure of incentives and justifications altered considerably with the introduction of the GLA, and particularly the Mayor of London as an actor. First, this actor did not have a specific incentive to protect economic development in any particular part of the city, as individual boroughs had. The Mayor was, therefore, able to take action across the city, action that would have been hampered by competitive concerns between the boroughs. Second, the electoral incentives for the Mayor—voted in by the popular vote across the whole of London—encourages the holder of the post to undertake policy action that will yield a demonstrable effect across the city within the electoral cycle. The current incumbent, Ken Livingstone, has identified transport and traffic as the key areas where he intends to have such an effect. Hence much of his early policy effort has gone into public transport, particularly the arrangements for the London underground system, and traffic reduction. Here congestion charging could act as a focal point for demonstrable policy action. Third, the London Mayor is relatively constrained in terms of powers and financial resources. Yet congestion charging is explicitly part of his allotted powers. Contemporary technology means that such charging would not only cover its costs but could also raise considerable sums of money, which would be an additional bonus for the Mayor. Finally, any electoral kickback from introducing a tax could be carefully managed by setting the charge so that it compares well with public transport fares (the main alternative) and providing appropriate exceptions to key groups (such as local residents and taxi drivers).

Furthermore, the discourse of congestion charging could fit within the dominant policy discourse that the Mayor was promoting. This focused heavily on the promotion of London as a world city. Air quality and the associated congestion became primarily an issue of *central London* congestion. The concern was the impact on international companies' location decisions, rather than local economic development. Thus, in the draft GLA London Air Quality Strategy (2001), there is a strong framing of the issue in terms of economic rationality, alongside the procedural rationality noted in the British Strategy (see Fig. 7.2). Right at the beginning, the point is made that 'Poor air quality provides a disincentive for businesses to create jobs in London', particularly for international companies (ibid. 2). The key challenge is creating 'sustainable growth' (ibid. 3).

In pursuit of this goal a number of solutions are presented, falling into three main categories. First, there is a heavy emphasis on technological solutions, although these largely fall outside the Mayor's competence and the strategy is

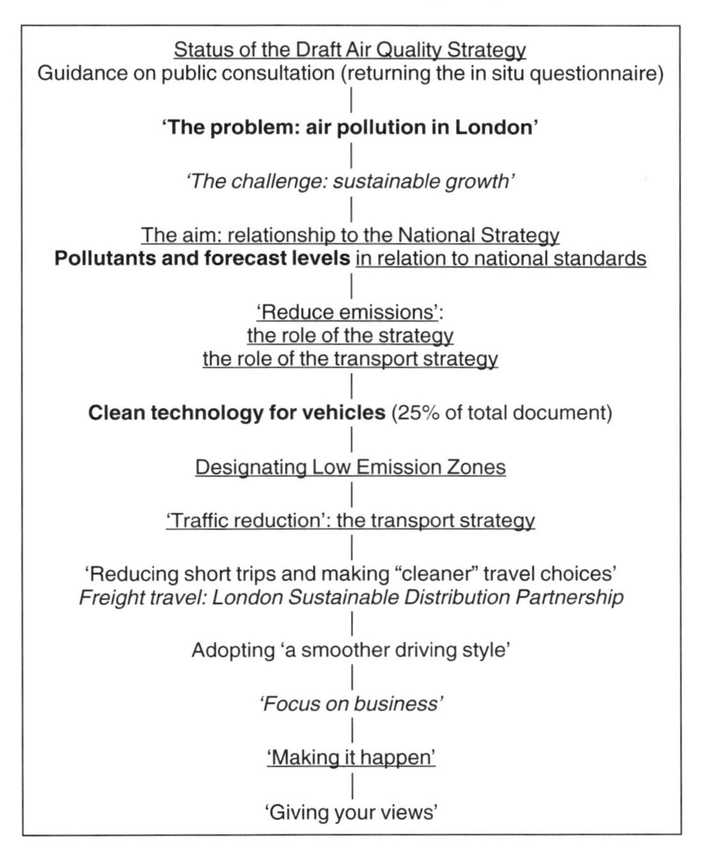

Fig. 7.2. A rhetoric line for the London Air Quality Strategy, *Clean Air for London: Highlights of the Mayor's Draft Air Quality Strategy*

Note: Underline indicates the rational policy process discourse; **bold** indicates the discourse of scientific rationality; *italics* indicates the economic rationality discourse; plain text indicates the communicative rationality discourse.

reduced to encouragement and publicity in most cases. These technological solutions hark back to an applied form of scientific rationality again but, in this case, they are heavily constrained by the arguments of economic rationality. Solutions do exist in the form of fuel-cell cars, particulate traps, etc., but their implementation is constrained by the costs of modifying the motor vehicle fleet and the economic dynamics driving research and development in this area. Second, there is the reliance on procedural rationality by emphasizing the links with the Mayor's Transport Strategy. It is here that congestion charging features, although it actually plays a relatively minor role in the air quality strategy. It features more prominently in the Transport Strategy, as a congestion reduction measure with associated benefits for promoting London as an international location. Third, there is an emphasis on encouraging behavioural change at the level of the individual, family, and (to a lesser extent) firm. This is where air pollution policy connects with communicative rationality.

Communicative Rationality and Air Pollution

Communicative rationality does not feature particularly within policy discourses surrounding either air pollution control or air quality management. Within procedural rationality there is usually a place for consultation exercises with the general public. But this does not really fulfil the basic requirements of communicative rationality. Beyond this, most pollution discourses limit their engagement with the public and community groups to the provision of information.

The requirement to provide public information is recognized within IPPC, with provision for public registers of authorizations and breaches but also a more recent provision for a register or inventory of polluting emissions. While there is a general commitment for this information to be public and available (following EU directives on environmental information) there are in fact two major categories of permitted omissions. These are on the grounds of national security and commercial confidentiality, the latter demonstrating again the influence of economic rationality in this domain. If a commercial advantage is reduced by the release of information, or an unreasonable commercial disadvantage produced, then the information may be withheld. This provision of information supports the notion of a transparent and accountable regulatory process and holds out the prospect of the 'soft' sanction of shame encouraging norms of compliance within the industrial sector. However, it is not supportive of a more generalized discourse involving all parties on the issue of air pollution. If an environmental group, such as Friends of the Earth say, uses this public information to discover poor environmental performance on the part of a particular company, it can publicize this and demand that the Environment Agency take enforcement action, but the NGO itself is not empowered within policy debates.

The provision of public information in the case of air quality management plays a different role. As Bickerstaff and Walker (1999, see also forthcoming) identify, there are three reasons for providing information to the general public about air quality: that certain people need to know in order to safeguard their health; that people have a right to know the state of air quality; and that knowledge can provide the basis for people to take action to reduce pollution. The idea of rendering policy more transparent by such information provision is still relevant, but there is the additional emphasis on behavioural change for private or public benefit. The general importance of information in encouraging changed behaviour is identified in the EU's sixth Environmental Action Programme, where the following is recommended to ease the implementation of environmental legislation:

To empower citizens and change behaviour, the following actions are suggested:

- Helping citizens to benchmark and to improve their environmental performance;
- Improving the quality of information on the environment.

(Section 6)

But in the case of air quality management, the emphasis placed on information provision can support a shift in policy responsibility. It is a response to a situation in which more proactive forms of state action (through regulation or direct investment) are recognized as difficult and often deferred. The resulting emphasis on monitoring and modelling has been noted and, while this is being developed, the solution comes down to providing information on air quality so that individuals can take action themselves; if they then suffer from poor air quality or pollution, it must be their fault. Even where other policy actions are being pursued, as with the GLA draft London Air Quality Strategy (2001: 15), an emphasis on individual behavioural change persists: 'The choices we make on how we travel affect the quality of air we breathe. If all Londoners made an effort to cut back the number of short trips they made by car, London's air quality would be improved'.

However, this approach has significant limitations. In their research on Birmingham, Bickerstaff and Walker (1999) found that air quality information was seen as complex and difficult to make sense of, couched in ambiguous terms, open to challenge on the grounds of accuracy and validity and, finally, not always relevant at the local scale. They found that, far from relying on the claims of scientific rationality bound up in this information, people used their lay knowledge and experience of air quality both in preference to official information and as a benchmark against which to judge that information. They argue (ibid. 291–1) that this provides support for: 'a model of environmental communication which emphasizes not the deficit of public understanding, but the active interpretation, judgement and evaluation of official information sources through processes of contextualisation and in relation to the "real world" of day-to-day encounters.' A similar reliance on lay knowledge was found in research on air pollution in Lyons (Coanus *et al.* 1998). Bickerstaff and Walker (forthcoming) have also found that such a shift of responsibility, or laying of blame, is resisted by the public and blame is seen in much more complex and nuanced ways.

Consequently, where difficult air pollution problems lead policy-makers to try and shift some responsibility onto the general public, it would seem insufficient just to provide information and expect behavioural change. Rather policy-makers will need to engage with the very real dynamics that underlie the claims of communicative rationality. These are very likely to challenge both scientific rationality—in terms of distrust of the scientific information—and procedural rationality—in terms of distrust of the state itself due to its lack of effective management of the very problem defined by the air quality information. But currently it is a very attenuated form of communicative rationality that is operating in air quality management policy debates. There is no need to involve the public or other key stakeholders in order to legitimize the policy. The claims of scientific, economic, and procedural rationality suffice to justify policy action. It should follow expert leads in identifying the key pollutants; remain within the bounds of reasonable economic costs; and be rational in its policy procedures. The public has a right to know about the policy and its outcomes but involvement does not need to go beyond this.

Indeed, given the very diffuse nature of the environmental impact involved, it will be extremely difficult to involve the general public. The collective action problem in the case of a public good such as a high-quality atmosphere is amongst the most severe. So in what kind of situations might a stronger communicative rationale hold sway? One key example of a successful campaign on diffuse air pollution and associated air quality is the British anti-leaded petrol protest of the 1970s focused around the NGO CLEAR. This led to the introduction of a differential petrol duty for unleaded petrol and eventually to the withdrawal of leaded petrol at the end of the 1990s. What were the key factors here?

First, there was an increasingly strong scientific case linking lead in the atmosphere with harmful effects on children; particularly on their intellectual development. Second, economic rationality became a decreasingly important reason for not taking action, since technological development meant that leaded petrol held a less and less significant market share. Third, the relevant procedural rationality did not concern regulation in the first instance but the structure of the duties on different fuels. In this context it was relatively easy to adjust the structure incrementally, as indeed occurred on an annual basis at Budget time anyway. Fourth, the collective action problem facing the affected public was overcome for a sufficient group of people for them to exercise considerable political pressure on government. This was because the benefits of policy action were valued very highly by those affected, concerning as they did the health of their children. And since the health impacts were identified as particularly concentrated in areas adjoining busy roads (particularly schools), this enabled specific, delimited communities to identify themselves as key beneficiaries of policy action. Such communities were relatively easier to mobilize, to maintain at relatively high levels of collective action, and to monitor for non-involvement, at least to some degree.

In many cases, though, there is insufficient institutional basis for extensive reliance on communicative rationality within air pollution and air quality management policy. To create the opportunity for this would require institutional change. For example, there is a possibility within future European policy developments of opening up air pollution policy to communicative rationality. This lies with the proposals for environmental liability that are currently being considered by the European Commission. As well as requiring polluters to be directly, legally, and financially responsible for the environmental harm they cause, these proposals are currently considering giving public-interest groups the right to require government bodies to act in these matters and also to challenge government decisions directly in the courts. This means that environmental NGOs and affected communities would be given legal standing within pollution control policy. It envisages a role for such organizations to oversee the role of environmental agencies, as well as act as watchdog for companies who might pollute. This would considerably expand the role of NGOs and communities within the pollution policy domain and might provide organizational sites as well as an institutional justification for an extension of communicative rationality within environmental planning.

Or it may be that other institutional arrangements provide opportunities that communities can take advantage of. An interesting example of community-level action is provided by the Clean Air Communities initiative in the US (www.cleanaircommunities.org, accessed 11 June 2002). This is an NGO operating in the context of market-based mechanisms for trading reductions in emissions to air. The organization develops projects that will reduce emissions and then benefits financially from trading these emission credits on the credits market. Such a potential for raising funds provides a strong incentive for overcoming collective action problems on air pollution and has been effective in creating community-level action to implement new technologies. The institutions of market-based instruments, while rooted in economic rationality, are supporting community activism. This provides further evidence of how revised institutional arrangements are the key to furthering environmental protection and providing the context in which policy for improved air quality can be legitimized, in this case by a combination of economic and communicative rationality.

Conclusion

The case study of air pollution control and air quality management highlights the importance of scientific rationality within environmental planning, but it also shows how this connects with procedural and economic rationalities. There is much more limited scope in this case for drawing on communicative rationality, although revised institutional arrangements could provide opportunities for extending this rationality within the air pollution domain.

8

Housing Land Policy

One of the key tasks that environmental planning is charged with concerns the allocation of land for residential development, that is, the formulation and implementation of housing land policy. This task provides a good case study from the institutional discourse approach because it invokes three of the four rationalities discussed above. First, it provides a good example of a particular form of procedural rationality; in the UK there are well-developed procedures in place for allocating development land. Second, the involvement of key economic interests in the form of residential developers introduces economic rationality into these debates. In the UK there has also been sustained involvement of these economic interests in the procedures of planning, resulting in an interaction of procedural and economic rationalities. Third, the allocation of new development continues to be a highly politicized matter, often resulting in vociferous collective action on the part of local communities. Hence communicative rationality and the involvement of community voices within the planning process are relevant issues. The interaction of these three rationalities provides the central material for this case study. In addition, housing land policy has recently become a focal point for attempts to mainstream sustainability concerns by incorporating them within procedures and decision-making. This evolution of a sustainability discourse within environmental planning will also be discussed.

Procedural Rationality and Housing Land Policy

Land-use planning systems (also known as urban and regional, or town and country planning systems) are essentially procedural in nature. The system establishes the stages and routines for generating plans, which then guide the allocation of land for development or directly constitute that allocation (in the case of zoning systems found, for example, in the USA). In Britain, land is allocated in a plan for residential development but actual permission to develop land is granted through development control, the consideration of proposals submitted as planning applications. There are, therefore, two elements to the planning system. But, as decision-making within development control has to have regard to the development plans, development planning will be the key focus here.

Currently development plans set out, in more or less precise map form, the areas in a locality that are to be developed for housing. Due to the confused and confusing local government reorganization carried out in England in the 1990s, there is a distinction between parts of the country where there is a two-tier system of local government and those with a unitary system (as in Scotland and Wales). In two-tier areas there are county and district councils. Counties prepare more general or strategic Structure Plans; districts prepare more detailed and spatially precise Local Plans. In unitary areas, a unitary development plan is prepared with a more strategic part—Part 1—and a more detailed part—Part 2. The idea, in either case, is that there is a hierarchy with the more detailed plan fitting into the framework established by the more strategic one.

Above this level of development planning sits the regional level. However, in England there is currently no regional tier of government. Therefore Regional Planning Guidance is prepared by associations of local authorities known as regional planning conferences or, in some regions where more formal institutions have been put in place, Regional Chambers. These bodies prepare general guidance on the allocation of development across their region. In turn Regional Planning Guidance sits beneath National Planning Policy Guidance, prepared by central government and currently comprising 25 Guidance Notes. These are mostly topic-specific and include one on housing, although many of the other topics also impinge on housing land allocations (such as the transport, nature conservation, and flood risk notes, to mention just three).

This system is currently the subject of proposals for change, as set out in the 2001 Green Paper on Planning *Planning: Delivering a Fundamental Change*. The proposals seek to streamline the system. In relation to housing land policy, this will largely take the form of abolishing the Structure Plan and replacing the Local Plan (or the Unitary Development Plan) with a Development Framework, which is more exclusively focused on allocating development land than current development plans. The rationale for this change lies partly in another recent policy innovation, the introduction of Community Strategies prepared by local authorities along with other key local stakeholders working together as a Local Strategic Partnership. Since this Community Strategy is supposed to cover all issues of concern to a locality, the idea is that a more specific Development Framework can fit within this and highlight the land allocation implications.

The other proposed change of relevance here concerns the regional level, where Regional Planning Guidance will be replaced by Regional Spatial Strategies, again more closely focused on the land allocation implications of the regional view. The institutional arrangements for preparing these Regional Spatial Strategies have yet to be confirmed but it is likely that a regional partnership of key actors will be involved, including the relatively recently established Regional Development Agencies, which focus on local economic development.

This brief discussion of the organizations and arrangements involved in development planning establishes just the bare bones of how housing land policy works. It highlights the way in which land allocation (in both the current and proposed formulations) is addressed repeatedly on different spatial scales with different degrees of precision, leading down eventually to a detailed mapping. What it does not illustrate is the nature of the assumptions, routines, and working practices that actually link these different tiers together and how these also bind actors together into a specific housing land discourse. In the case of housing land policy, the key aspect of this discourse is the transformation of policy into numbers and then into spatial allocations. It is through these transformations that population forecasts are tied into land allocations, and the idea of providing housing for people (in aggregate) is transformed into the idea of providing land for housing development (on specific sites).

The starting point for this transformation is provided by national population predictions, prepared by the central government Office for National Statistics. These are recalibrated as household numbers, then as dwelling requirements, and finally as land requirements. The initial policy problem is change in population but the end result has to be an allocation of land that can be presented in a development plan and then inform regulatory practice in development control. Murdoch, Abrams, and Marsden (2000: 203) have explored how this transformation works in practice and how it gives rise to a particular technocratic form of procedural rationality. As they put it: 'In short, the whole structure is held together by numbers. . . . the figures flow down the hierarchy from the DoE [Department of the Environment, now Department of Transport, Local Government, and the Regions] through the regions, counties and districts as dwelling projections to be incorporated into plans and, finally, emerge as buildings in the landscape.' In this discourse of numbers, there is a clear appeal to procedural rationality at work. If the calculation of the numbers and their transformation are conducted properly, then a rational and optimal outcome in terms of these 'buildings in the landscape' will result.

The suggestion is that it is a simple technical exercise to recalibrate by applying appropriate multipliers: if the population number is x and average household size is y, then predicted households are x/y; if one allows for a % of one-person households, b % of two-person households, etc., then one can work out the dwelling requirements; and at densities of k dwellings per hectare this will work out at a certain acreage for land allocations. However, these transformations involve many assumptions. There is the assumed accuracy of the population predictions, the assumptions about how the population is structured into households, how it will migrate (or not) around the country, what kind of housing they will need or demand (via the market), and the preferred spatial location of that housing. The numerical transformations incorporate these assumptions and often hide them, presenting them as technical activities rather than value-laden working practices.

The presentation of housing land allocation as a technocratic exercise in this way has two implications. First, it legitimizes the bureaucrats who engage in such exercises and elevates the significance of their working practices, labelling them as technical and holding a form of expertise. It is thus in bureaucrats' interests to support and use such a technocratic numbers discourse. It can help insulate them from criticism in what is a highly politicized policy area. But, second, it cannot dispense with the politics of housing land allocation; this is intrinsic to the impact of housing land allocations on specific interests. Therefore this technocratic discourse itself raises the issue of how these different interests are dealt with within housing land policy and whether this discursive formation has to contend and compete with other discourses. In the case of housing land policy, the role of the economic interests—the housebuilders—is central.

Economic Rationality and Housing Land Policy

Housebuilders are central to housing land policy because planning for residential development cannot have regard to procedural rationality alone. The ability to provide these buildings does not, for the most part, rest in the state's hands. In Britain, the vast majority of new housing is built by private-sector housebuilders for sale to owner-occupiers in order to generate profit. This is, therefore, a clear example of public policy having to take account of the corporate discretion of the commercial sector (Lindblom 1977). It has been shown that this gives housebuilders a specific form of privileged access to policy-making at central and local levels (Rydin 1986). But it also means that the discourse of housing land policy has to take account of economic rationality along with this procedural rationality. The economic actors, the housebuilders, are the only ones with the practical economic expertise to fill the gaps in this numbers game.

For the transformation of population figures into dwelling figures and related land requirements only goes so far. The planning system requires that specific sites are identified, either in the plan itself or through the development control process as developers identify new sites (sometimes referred to as 'windfall' sites). This raises a range of issues about the ways in which specific sites can or cannot be developed, particularly within a specified timeframe. What are the barriers to site development? Who owns the land? How much of the site can be developed? What is the appropriate density? What kind of housing, in terms of size, style, and design is appropriate for this site? The answers to these questions mix technical issues (such as ground condition) with policy issues (such as acceptable design styles, parking norms, and the safety of the road layout) and economic issues (such as the kind of housing that local consumers demand). On these economic considerations in particular, planners find themselves having to defer to the experience of housebuilders.

As Murdoch *et al.* (ibid. 209–10) spell out:

> the whole debate around the levels of development revolved around the figures. Even though a well-organized and long-standing anti-growth coalition was deeply suspicious of the 'numbers game' that they were forced to play, they could find no way to resist the technical calculations coming down from 'on high'. The only actors with the required technical knowledge to play the game successfully were the house builders. . . . Thus, the configuration of local argument by the national-to-local calculations of housing demand enabled the house builders to dominate the debate.

In this way, housebuilders' experience is reframed as expertise, that is expertise relating to judgements about which sites can be developed, in what order, at what density, and in what time-frame.

The institutional context in which these judgements are expressed and economic rationality relied on, is the housing land study. This refers to an assessment of sites in a locality, using the local housebuilders' economic expertise to ratify (or discard) the allocations identified by local planners. The planners identify local sites which, together, could be developed to produce a given number of dwellings. This number should match up with the regional, county, and district breakdowns of dwelling requirements, derived from the population predictions. These sites are then assessed by the housebuilders, leading to a negotiation about which sites are 'available' for development and which are not. The result is an agreed list of sites that can go forward into the development plan as land allocations. After initial resistance when such land studies were introduced in the early 1980s, this discourse of practical economic expertise is well established within housing land policy and is sufficiently taken for granted to be expressed completely overtly in most cases. It is accepted as a legitimation for the selection of sites for residential development.

However, to reinforce its dominance discursively, the discourse of practical economic expertise retains strong links with the theoretical economic discourse. In effect, it mimics economic theory, although within a framework provided by procedural rationality. Although the demographic targets, forecasts, or predictions are provided by governmental statisticians through a set of institutionally accepted procedures for extrapolating past population trends and dividing national figures among regions, they are here recast as a form of housing demand. Of course, in neo-classical theory, demand implies ability to pay as well as certain preferences. The population figures by themselves carry no such implications. By turning them into housing demand, all sorts of assumptions about preferences and financial resources are sidestepped. A particular form of household structure is normalized. This turns population structure into housing preferences and the housing market is then seen as the route to meeting these preferences, satisfying this demand.

Satisfaction of demand is provided via supply in the neoclassical model. So land availability is recast as housing land supply. As indicated above, local builders largely identify this, using their local knowledge of land ownership, site constraints, and market conditions. Housebuilders are, therefore, using their knowledge of market conditions (including market demand) to determine land availability. For this reason, land availability cannot be considered as a true surrogate for the supply function of the neoclassical model; this would be based on the cost functions facing housebuilding firms. Furthermore, in the neoclassical model, supply and demand would be independent. Given the way that housebuilders assess housing sites (using demand conditions among other factors), this is clearly not the case. But this is not important in the discursive domain of policy; the fact that the land availability assessment mimics the theoretical economic discourse is important. This adds to the assessment's authority, using the theoretical discourse to support and re-present the discourse of corporate expertise within new housing production.

Communicative Rationality and Housing Land Policy

This discursive structure also has implications for the potential of communicative rationality to penetrate policy discussions. The key issue here is how a policy discourse, heavily framed in terms of economic and procedural rationality, can cope with the introduction of other voices and claim the rationality of open dialogue oriented towards mutual understanding. This can be explored at two levels: the more general cultural debate over housing land and the site-specific interaction between this housing land discourse and collective action within planning contexts. In each case, it is notable how the involvement of other voices within housing land policy does not open up policy to communicative rationality, but rather produces a conflict between anti- and pro-development arguments.

At the level of general cultural reference points, the discussion around housing land begins from the numbers game and then is organized either in support or against. Dialogue does not generally proceed beyond this, to the voicing of alternative positions or the proposition of compromises. Rather, it becomes a repeated voicing of these two polar positions. Examples of this are particularly found in the media, where this kind of conflict particularly suits the routines and norms of news production (see Ch. 1). To explore this further, two typical examples will be briefly analysed: an article by Paul Barker in *The Independent* and one by Simon Jenkins in *The Times*.

Paul Barker's article is summarized in Fig. 8.1 This shows the heavy reliance on the numbers games as seen through the lens of economic rationality. The central argumentation follows neoclassical lines, namely that house prices will fall (or at least level off) if more houses are built. This point is repeatedly made, but while the article actually draws on a research report for the Joseph Rowntree Foundation, the main support is found in the views of developers, particularly as concerning which sites are in the right place and are commercially viable. Market processes are seen not only as providing the key reference point for appropriate housebuilding, but also as underpinning the natural

Fig. 8.1. A rhetoric line for Paul Barker's article

Source: *The Independent* 10 March 2002.

population geography of Britain. If people want to live in a South-East that is 'bursting at the seams', they should be allowed to do so; it will make them happy. Happiness, preferences, and market demand all shade into one another. Resisting these patterns is 'social engineering', based on taboos and electorally unpopular. This expresses a strong anti-planning sentiment: the Government should 'stop sitting on people's aspirations'. And yet this same government is charged with getting more land to market.

In sharp contrast to the next article considered, there is very limited drawing on general cultural reference points. Another article on the Joseph Rowntree Foundation report (in the *Guardian*, 4 March 2002) quoted the report as rec-

Fig. 8.2. A rhetoric line for Simon Jenkins's article

Source: *The Times* 8 March 2000.

ommending throwing off 'the folk memory of the slum clearance disasters of the 1950s and 1960s'. But no cultural resources were offered as a replacement for this negative memory, other than an appeal to the Labour and Conservative Parties' memories of their role in past planning for residential development. This form of argumentation simply reinforces the numbers game, its central role in housing land policy and its connections with economic rationality. It then becomes the key reference point for contra-argumentation.

A good example of anti-numbers game rhetoric is provided by the well-known journalist Simon Jenkins in his article in *The Times* (summarized in Fig. 8.2), which criticizes the housing land allocations announced by central government in early 2000. In these allocations, two themes were emphasized. First, there was the absolute need to provide millions of extra houses over the next fifteen years in order to cope with demographic change (originally 4.4 million, then downgraded to just under 4 million); second, there was the announcement that 60 per cent of these houses should be built on 'brownfield' land.

'Brownfield' is itself an interesting discursive term, used to contrast with 'greenfield' and to suggest something dirty and used. Guidance on how this term should actually be translated into site allocations has provoked some rather tendentious discussion. Brownfield is now generally used to mean land that has previously been developed but, as urban wildlife interests have pointed out, this does not mean that it does not have biodiversity value; numerous such sites act as habitats to more flora and fauna than many an agricultural field.

Jenkins's article provides several interesting rhetorical features. Although the rhetoric-line does not fully reveal this, the language of the article is highly emotionally charged. The reference to the 'rape' of the countryside is one indicator of this; elsewhere there is talk of 'a despairing public' screaming, of households as 'refugee armies', and of downtowns 'left to rot'. This angry ethos is used to suggest a reasonable person pushed to the limits. The diatribe format is then used to challenge both the procedural and economic rationalities that constitute, as shown above, the numbers game. A government report by the former Chief Planning Inspector is denounced as 'gibberish' but, more significantly, procedural rationality is equated with wartime planning and with socialism; the Secretary of State is envisaged as 'brooding daily under a bust of Lenin'. Meanwhile economic rationality is acknowledged (as in the acceptance that farmers will readily sell land for development) but the role of government is seen as protecting 'what the market will not protect'. The fault of government policy is thus couched as being out of touch with public values.

The epitome of these values is a feeling for the countryside: 'British public opinion pleads for the protection of the countryside'. Here there is the familiar elision of 'feeling for the countryside' with nationhood and patriotism. The rhetorical connection between rurality and national identity has also been identified in analysis of green-belt discourse and the long-lived success of this discourse. A complex connection of concerns about the city and urban unrest are allied with this emphasis on rural areas as the metonymy of Englishness (Rydin and Myerson 1989). At this point Jenkins indulges in the common discursive strategy of the personal testimony to a rural past: 'the Surrey village in which I grew up'. Here the ethos shifts to emphasize personal experience and the authenticity of that engagement with the countryside. This links back to the romantic discourse, discussed in relation to Lovejoy's and the Prince of Wales's Reith lectures in Ch. 6. What is notable though is that—for all these references to socialism, Englishness, and romanticism—the punchline is the demand for a complete government 'ban' on building in the countryside. Procedural rationality may be ridiculed but there is still a strong demand here on government regulation. Ironically this is the same punchline (or form of closure) as in the pro-development stories discussed above.

So, in the more general discussion of housing land policy, the numbers game heavily couched in terms of economic rationality is opposed by a relatively culturally rich but essentially anti-development rhetoric. Housing land policy debates become cast as pro- versus anti-development, numbers plus economic rationality versus a richer sense of community. This is as true at the site-specific as at the national level. Bridger (1996: 362) shows such a dynamic at work in his study of planning for residential development in Lancaster County, Pennsylvania. Here also 'Opponents of rapid growth and development wanted land-use planners to reflect what I shall call the agricultural narrative, while advocates of growth wanted a landscape that would accommodate the entrepreneurial spirit of residents embodied in the business narrative' or economic rationality. Quality of life was constructed in terms of the rurality of the area and its agrarian past, producing a distinctive and powerful metanarrative, which he labels 'rural mystique'. This made it seem 'eminently reasonable' to argue for greater regulation. However, he also noted that the 'principles of rational planning' were invoked to produce a win–win situation in which the interests of all parties to this policy debate were protected (ibid. 363). In this case, the cultural resources of an agrarian narrative were symbiotic with the procedural rationality of the planners.

This preservation of procedural rationality is also apparent in Murdoch, Abrams, and Marsden's work. They tell of how the interaction of interest networks in a specific location resulted in a small increase in housing land allocations, this being dependent on the relative resources and power of the different actors within those networks. However, the broad framework and detailed argumentation of the plan remained largely unchanged, which highlights that the actual outcome of housing allocations depends on more than just the terms in which housing land is discussed. But the definition of that outcome as land allocations and the consequences of that definition (explored further below) are established by the structure of the housing land policy discourse. Murdoch *et al.* (1999) locate the reason for this in the closed discursive structure of the plan: 'its own story, its own momentum, its own modalities . . . It proved in the end to be effective precisely *because* it defied technical argument and could not be opened up so that the conditions of its production could be examined'.

In other words, the closed nature of the housing land debate in terms of a numbers game has become one of the taken-for-granted aspects of planning, one that is not challenged within planning circles or fully understood as shaping and constraining the debate. It is only challenged by an emotive anti-development rhetoric, which establishes the debate in terms of distinct oppositions and conflicts. Communicative rationality is difficult to establish in such a case, since a broader range of perspectives and views cannot find a place to be heard effectively within these debates. Hillier has argued it is the emphasis on technocratic expertise—as represented by the numbers game—that is disempowering for the range of other voices. She sees the key agencies taking 'refuge behind a discourse of instrumentalism and a technical system of "expertise" ' (1993: 107). Murdoch *et al.* (1999: 210) also claim that local actors 'simply could not resist the technical forms of expertise which bound their locality into a long chain of national-to-local regulation'. In their cases, local actors were able to overcome collective action problems and come together, but could then only engage in the policy process within the limited terms of debate offered by the numbers game.

It should not be forgotten that there is a discursive basis for collective action against the numbers game also. In earlier discussion (notably in Ch. 4) it was emphasized that, although the collective action problem was a general difficulty for local activism, it was most readily overcome in anti-development protests. Here the benefits of activism would fall on a geographically defined and readily identified group of people who saw considerable potential in lobbying local government in order to protect their highly valued residential assets and lifestyle. While often negatively stereotyped as NIMBY politics, this is nevertheless an important basis for active local politics on housing land issues.

So, to summarize, housing land policy is overwhelmingly framed in terms of a technocractic numbers game, which is defined in terms of procedural rationality but is strongly supported by economic rationality, both by drawing on the expertise of economic actors (housebuilders) and using the concepts of formal academic economic discourses. Where attempts are made to include other voices within the debates, in pursuit of communicative rationality, the dominance of the numbers game persists and constrains the debate. It further encourages other voices to fit within a simple dichotomy of anti- or pro-development views. There is a range of cultural reference points that can be used to support the anti-development viewpoint. Furthermore, an anti-development stance can help overcome collective action problems by reinforcing community identity and titling the balance between the costs and benefits of participation. This then further entrenches the limited terms of the housing land policy debate, within which local conflicts over housing land have to be played out.

Discourse, Interests, and Ecological Concerns in Housing Land Policy

Such an analysis of the discursive structure of housing land policy raises the question of who benefits (or loses out) as a result. This section considers how this dominant form of the housing land debate relates to planners, housebuilders, and communities as interest groups, and then raises the question of where this leaves ecological concerns.

Starting with the interests of planners, the dominant discourse is one that emphasizes the role of planners in 'allocating land'; the very phrase suggests a significant step in the process by which houses are produced. Research on the role of the planning system in relation to residential development has repeatedly shown how weak the planning system is in the face of the economic forces of housebuilding (Ball 1984; Rydin 1986; Healey *et al.* 1988; Shucksmith 1990). For allocating land in a plan is no assurance of that land being developed;

neither does it mean that other unallocated land will not be developed. The relative absence of any other policy tools—such as subsidies or tax-breaks, public landownership, or public housebuilding—means that the act of land allocation is not as effective as presented in its own policy discourse. Nevertheless, the discourse is significant in suggesting that responsibility for land allocation, and hence for residential development and the location of population, lies with the planning system and hence with planners. The housing land discourse supports the status of planners working with development plans but it can also make them the focus of responsibility and blame for outcomes (as seen in the media extracts at Figs. 8.1 and 8.2). Fortunately for planners, the benefits of status are felt personally and almost daily, the attribution of blame is only occasional and tends to fall on the profession as a whole.

The discourse also has significance because it focuses attention on certain aspects of residential development and ignores others. It has already been mentioned that the linkage from population numbers to household numbers to dwelling numbers and hence to land ignores many other aspects of the process by which housing is actually produced. Murdoch *et al.*'s analysis has highlighted how many assumptions, particularly about household size, migration between areas, density of development, and the amount of development on windfall sites (small sites which are not allocated in plans), are involved in the final decision about land allocations. The housing land policy discourse emphasizes the importance of land-use planning and hence planners in determining land allocations, and ignores many of the economic factors that determine housing production. In particular, this diverts attention away from the role of key economic actors such as housebuilders in producing residential development. Changes in the urban landscape become the responsibility of planners not builders. And yet it has been shown that housebuilders actually use the resources of economic rationality to justify their institutional role in filtering sites out of the land allocation process. They play a policy as well as a production role in terms of residential development. The numbers game obscures this for those viewing the planning system from the outside.

Finally, in terms of local communities, it has been shown how the discursive structure of the debate favours a conflict between pro- and anti-development stances and effectively encourages communities to adopt anti-development, NIMBY rhetoric. This can be extremely constraining in terms of the issues that communities may wish to raise, as well as limiting in terms of the success that communities may achieve in shaping planning decisions. Other arguments are beyond the bounds of acceptable or effective rhetoric. Only those communities who see their interests entirely in these terms will be satisfied. Furthermore, it encourages the identification of community in localized terms. This is part of the way in which collective action problems are overcome. But it also limits the extent to which a broader identification of community interests can be achieved. This is particularly pertinent to ecological concerns, where the emphasis is on impacts and causes relating to ecological systems that involve the local but go beyond a preoccupation only with the local. This problem of moving beyond a local NIMBY frame to a broader sustainability frame has been noted in some of the accounts of the environmental justice movement in the USA.

Indeed the numbers game of British housing land policy has a number of constraining features where ecological impacts are concerned. In the recent policy discussions about new housing and sustainability, attention has focused almost entirely on the implications for land allocations. The discussion has examined the assumed links between development density, the amount of road-based travel, and global warming. The dominant argument has been that higher density developments reduce the need to travel and hence travel-based greenhouse gas emissions. Because of the interactions between arguments about car-based transport and public transport within transport debates, the argument extended to a discussion about the location of housing land allocation in relation to transport infrastructure (see e.g. Planning Policy Guidance Notes). However, attention is focused away from the wide variety of other ways in which residential development can contribute to reduced carbon dioxide emissions and hence sustainability: notably, higher energy efficiency in construction and production methods for construction materials, as well as higher energy efficiency for the residential buildings themselves. Furthermore, the density-leads-to-sustainability line of reasoning ignores aspects of sustainability besides energy efficiency, such as water and habitat conservation where changes in other aspects of residential development could probably make a greater impact.

This policy case study shows clearly how the argumentative structure of the policy debate about housing land gives discursive opportunities and resources to some actors and denies them to others. This argumentative structure itself is a reflection of the need to legitimate planning processes and outcomes by reference to criteria of rationality, technical expertise, and objectivity. But it also reflects the interaction of the needs of the regulatory system with the plan-making system, both influencing and redefining each other (see Ch. 5) and the interaction of various actors in the regulatory process. The framing of the housing land issues as a 'numbers game' represents the attempt—largely successful—to contain all these demands and needs within a single construction of the problem. How this works out in terms of outcomes depends on the contingencies of the specific case.

However, it is clear that this discursive construction of housing land policy resonates with the professional interests of policy-making planners and is shaped by the needs of planners engaged in regulation. It has a logic that fits with procedural rationality but also connects to economic rationality and the location of housebuilders within the policy process. The potential of communicative rationality to incorporate a range of voices is also affected by its construction, which shapes the context for the interaction of interests within specific land allocation disputes. It also constrains policy options and, thereby, has material effects through the environmental and social impacts that actually result from residential development. To conclude this chapter, current attempts explicitly to incorporate sustainability concerns within housing land policy discourse are discussed.

Incorporating Sustainability Concerns

Given the limitations arising from the dominant housing land discourse, various attempts have been made to amend the operation of the key rationalities, that is procedural, economic, and communicative rationality. Within planning circles, the rise of the sustainability agenda more generally has prompted a range of different innovations, all procedural in nature. The key ones can be summarized as follows:

Environmental Assessment: Assessment of environmental impacts (which can include cultural impacts) arising from specific projects; emphasis on integrating multiple impacts.

Strategic Environmental Assessment: Assessment of environmental impacts of plans, programmes and policies.

Environmental Capital: Elements of the natural environment that have value and support activities (as physical, human, and financial capital do); it can be divided into a fourfold typology of: (constant and critical natural capital) × (material and post-material values).

Environmental Capacity: The capacity of an area to absorb further development or growth without unacceptable environmental damage and without undermining sustainable development.

Following (and to some extent anticipating) European Union directives, all major projects are now required to undergo environmental assessment and all plans, policies, and programmes are subject to strategic environmental assessment. This involves identifying the existence and significance of environmental impacts that are likely to arise from the project, plan, etc. This should then allow a judgement of the overall impact, taking the different impacts into account in an integrated manner. The other two concepts identified above—environmental capacity or environmental capital—are rather different in nature. They are not assessment procedures but concepts for identifying and potentially providing some kind of scalar measurement on valued and important aspects of the environment. As such, they can then feed into plan-making processes and regulatory decisions.

There has been a considerable expenditure of effort among both policy officials and NGOs on developing these different concepts and techniques (e.g. the two reports by CAG Consultants and Land Use Consultants for a consortium of agencies, 1997 and 2001). In part, this reflects the prevailing modernist ethos that still guides much environmental planning. As Edelman (1988: 56) points out, 'Techniques carry a strong appeal in a capitalist society that socialises people to see inventive methods as the avenue to a better life.' This kind of development work also tends to fall within the category of work that policy officials find most interesting and professionally rewarding. According to the argument of Ch. 5, they will, therefore, expand the opportunities for undertaking such work. But interestingly, NGOs have also put resources into such work (Grigson 1995; Jacobs 1997*b*; McLaren, Bullock, and Yousuf 1998). This reflects the view that influencing the practice of environmental planning and the very detail of how procedural rationality plays out is a way of steering planning more subtly towards the goal of sustainable development. Mittler (1999: 356) refers to these concepts as being 'a language which we must speak, if we want to move sustainability from being merely a woolly concept to something local politicians can both understand and implement'.

The difficulty with this argument is that moving sustainability closer to the language that politicians and planners speak can be seen as diluting the sustainability concept rather than radically altering planning practice and politics. It is notable that the introduction of environmental assessment procedures has not significantly altered the majority of planning decisions, although it may have altered some of the detail of development through negotiation and planning conditions. A further example of this dynamic at work can be seen with the concept of environmental capital.

Environmental capital (and its extension as 'quality of life capital'; CAG Consultants and Land Use Consultants 2001) is meant to inform policy practice by providing a basis for specifying the contribution of the environment. It uses the language of economic rationality ('capital') to highlight the way that the environment supports our social and economic processes but it also connects with scientific and communicative rationality. In this way, environmental capital is intended to act as a bridging point within the policy process to a number of different domains and ways of constructing the environment. Total environmental capital is divided two ways, into critical and constant capital, and into material and post-material aspects. Scientific rationality is invoked with regard to obtaining knowledge of material dimensions of capital and distinguishing critical capital (which cannot be used without breaching sustainability constraints) from constant capital. Planning should, therefore, be oriented towards protecting this critical material capital. But it is argued (for example, by Owens 1994) that planning should also protect critical capital with post-material value, such as cultural heritage sites and landscape values. Scientific rationality has little to offer here, so instead communicative rationality is invoked to value these post-material dimensions. Practically, some form of community consultation is used to arrive at these values.

In the case of environmental capacity, such thinking has been taken even further. While the language of environmental capacity carries with it strong overtones of scientific rationality, deriving from ecological science and its concept of carrying capacity, Jacobs makes it quite clear that there is no scientific basis for specifying or even identifying capacity in most situations of human engagement with the environment. The idea of global capacity constraints on human activity from a sustainability perspective informs this concept, but the link with scientific rationality rapidly becomes lost as the scale moves from the global to the local, from issues of climate change to ones of local culture and way of life. Here there are choices to be made about how global capacity is divided up spatially and how to value aspects defined as non-material. Hence Jacobs (1997*a*) calls for a deliberative process by which communities may together socially construct the concept in a way that reflects their own values. These points are now also being made in relation to 'quality of life capital'.

The first point to note about all these concepts is that they have been devised with the intention of informing policy practice and, indeed, most have been refined by policy actors within local councils (BDP 1994; West Sussex County Council 1996), central government (DETR 1997) or quasi-governmental agencies (such as English Heritage, English Nature, Environment Agency). These are intended to be practical concepts; their expression, therefore, is couched in terms of the methodology for applying them. In some cases, such as environmental capacity and environmental capital, the methodology is relatively open-ended; there is explicit scope for variation in terms of what is included and taken into account. This is essentially a technocratic discourse, developing new elements of jargon and associated policy practices. There is a strong emphasis on measurement, on turning the valuation of the environment or the assessment of demands on the environment into a number, cardinal or ordinal. The echoes of environmental valuation are deliberate. The intention of creating this numerate language of sustainable development is to replace the market-driven monetization of the environment associated with economic rationality. Hence the emphasis on numbers is in the hope that one number can be replaced with another, one technique with another, one rationality with another, within the policy process. Sustainable development and economic rationalities mirror each other at this point, in order to induce a sleight of hand.

There is a clear danger here that the emphasis on methodology and technique will result in the new concepts being subsumed by procedural rationality and, as a result, add nothing new. Thus Connell (1999: 10), drawing on his professional experience in West Sussex, describes the environmental capacity approach as 'versatile and able to be absorbed readily into other processes'. The same point applies to various new procedures that have been adopted, ostensibly to further environmental protection or move towards sustainable development. For example, T. Richardson's (1999) study of strategic environmental assessment and local transport planning shows how strategic environmental assessment (SEA) readily became part of the prevailing procedural rationality. Rather than producing a new policy culture, SEA helped to institutionalize a particular form of technical 'policy knowledge'. The same point could be made about all the other procedures identified above.

The second danger is that the concepts become just a way of appearing to handle established political conflicts without resulting in any new resolution or outcome that is more favourable towards sustainable development. Connell clearly sees one of the attractions of concepts such as environmental capital and capacity as enabling planners to cope in 'the increasingly politicised context of limits of acceptable change' and he refers to them as 'a technical solution to a political problem' (ibid. 1–2). Particularly because of the way that these concepts link into the 'familiar discourse of limits', there is the very real possibility that they will become just another tool in NIMBY-type campaigns (Rydin 1998*a*). Another aspect is the way that such concepts and procedures can exclude certain actors. In his research, Richardson noted that the policy process was lacking in transparency and that the policy analyst kept control of the SEA process, acting in an exclusionary manner. Consequently lay knowledges were marginalized and policy knowledge used in preference to environmental or local cultural knowledge.

The emphasis is therefore thrown back on the other key rationalities: economic and communicative. The idea of incorporating ecological and sustainability concerns more directly into economic rationality invokes the hope of the ecological modernization discourse, that a form of green economic development is possible in which technological development will be used to reduce significantly the environmental impact of economic activities. Hence the prospect held out is of a green housebuilding industry. There are indeed some moves in this direction but they have been severely limited (Bhatti 1994). The evidence for a green economic rationality of residential development is scant. There are three institutional constraints on the spread of such a discourse. First, this is an industry resistant to technological innovation, with such weak methods of management control that previous attempts at innovation have often proved disastrous (Ball 1984). Second, it is an industry that largely operates by reaping economies of scale through avoiding tailor-made design solutions in favour of a pattern-book approach. This reduces the impact of recent green architectural innovations since the routine use of existing design practice is preferred over the potentially problematic implementation of the new. Third, the industry perceives consumer demand as highly conservative and tends to avoid innovations in the product—the dwelling—unless clear advantages to the consumer, as in reduced heating bills arising from greater energy efficiency, are demonstrable. Therefore attempts to green housing production have had to rely on increased regulation, primarily through the Building Regulations, although Shove and Guy's (2000) study has shown how the process of constructing these regulations are themselves the subject of complex and contradictory processes.

So what are the prospects for an ecologically aware communicative rationality? This will be explored further in the next chapter, where the case of Local Agenda 21 will be considered, since Local Agenda 21 specifically seeks to develop ecological awareness on the basis of communicative rationality.

Conclusion

The housing land policy case study shows the way in which economic rationality and procedural rationality can reinforce each other in the institutionalized numbers game. It identifies the problems this can pose for justification in terms of communicative rationality, particularly where the incentive structures are a strong influence on the collective action around housing land issues, reinforcing the terms of the debate as pro- or anti-development. It has also demonstrated the attempts to incorporate ecological concerns into housing land policy and the problems that this faces, with procedural rationality tending to subvert the distinctive justification offered by ecological concerns based in scientific rationality.

9

Local Agenda 21

Local Agenda 21 traces its history back to the Rio Summit and UN World Conference on Environment and Development held in 1992. At this event a document known as Agenda 21 was presented. It represented a manifesto for sustainable development for the twenty-first century and it received considerable attention and support at the conference. Many governments announced a willingness to adopt A21 and follow it up at the national level. A UN Commission on Sustainable Development has, since then, sought to hold governments to their word. Within Agenda 21, one chapter—drawn up with the involvement of local government organizations in the immediate run-up to Rio—sought to identify the importance of the community level to sustainable development. It encourages all local communities (a term largely interpreted in terms of local governments) to engage in their own Agenda 21 process, to develop their own local manifesto for sustainable development, to undertake a Local Agenda 21.

Local Agenda 21 (LA21) has proved to be highly attractive to local governments and communities across the world. There has been a considerable amount of policy effort directed at such local action, detailed in numerous case studies reported in journals such as *Local Environment* and in *Urbanisation and Environment* (see the special issues in 1998, 1999, and 2000) and in the volume edited by Gilbert *et al.* (1996) prepared for the UN Habitat Conference in Turkey. Research on LA21 is now able to detail and analyse the nature of this activity (e.g. Voisey *et al.* 1996; Lafferty and Eckerberg 1998; Lafferty 2001; Young 2000; LASALA 2001). The significance of this LA21 activity is the central place that commmunicative rationality holds in its justification, which will, therefore, be the starting point for the analysis.

Communicative Rationality and Local Agenda 21

An integral aspect of the LA21 idea was that an innovative approach would be taken, which would distinguish this new activity from prevailing local government practice, thus reflecting the innovative character of the concept of sustainable development itself. Such a goal required a new way of carrying out policy and even a new way of reaching agreement on that policy goal. For both these reasons LA21 is seen as implying a new mode of environmental governance, involving a broader and deeper engagement of all key local 'stakeholders', which—given the comprehensiveness of the concept of sustainable development—means representatives of all sections of society, including business and government alongside local communities and social groupings. Here the clear link to and reliance on communicative rationality is established.

This commonly leads to a distinction being made between the process of LA21 and its outcomes (whether procedural, in terms of a plan, or substantive, in terms of impact on the environment or sustainable development). Substantial emphasis has been laid on the process dimension. Early analyses have all repeated the mantra that LA21 was 'a process and not a plan' (Aygeman and Evans 1994; Selman and Parker 1999), a process seen in terms of public participation and education but more than that, as empowerment. Research has pointed to the variety of new techniques for involving local communities that have been adopted: visioning workshops, working groups, focus groups, round tables, village appraisals, 'planning for real' exercises (Young 1996; Burgess, Harrison, and Filius 1998; Roberts 2000). And while some of this research might be criticized for selective sampling and an overemphasis on 'progressive' authorities, the overall trend seems to be a modest shift towards more participatory methods being adopted by local municipalities (Selman 1998; Wild and Marshall 1999; LASALA 2001).

Within the documentation surrounding LA21, the language of communicative rationality can be readily found. A quote from ch. 28 of Agenda 21, which outlined the principles of LA21, provides a central example of this:

> 28.3 Each local authority should enter into a dialogue with its citizens, local organizations and private enterprises and 'adopt a Local Agenda 21'. Through consultation and consensus-building, local authorities would learn from citizens and from local, civic, community, business and industrial organizations and acquire the information needed for formulating the best strategies. The process of consultation would increase household awareness of sustainable development issues.

The key elements of communicative rationality are found in the reference to 'dialogue', 'consensus-building', and local authorities learning and households gaining 'awareness'.

The implications of such a dialogue are followed through in the UNED Forum book *Multi-Stakeholder Processes for Governance and Sustainability: Beyond Deadlock and Conflict* (Hemmati 2001), which looks to the involvement of diverse actors within joint debating and decision-making forums to provide policy solutions towards sustainable development:

> The increasing popularity of group-based decision-making reflects a widely shared belief that group decision-making offers the potential to achieve outcomes that could not be achieved by individuals working in isolation. (p. 74)

> Multi-stakeholder processes have a great potential to assemble, transform, multiply and spread necessary knowledge and to reach implementable solutions. (p. 94)

Such approaches are now seen as the necessary and logical consequence of adopting the goal of sustainable development.

A further example of the use of communicative rationality is provided by the Environment Agency's guide to *Consensus Building for Sustainable Development* (1998). This was written to guide the Agency's offices in their involvement in A21 and LA21 processes, particularly through their Local Environment Agency Plans. As the title suggests, this places consensus building as the central concept and conceives of this as '"primarily" bottom-up, involving initial options as well as preferred proposals'. It further argues, 'In reality, consensus building can take many forms and have a range of purposes in sustainable development decision making; the most important characteristic of its use is the spirit in which the process is adopted.' Such language is found in local LA21 documents also. To demonstrate this most significant level of LA21 discourse, a specific example will be analysed. This is necessary because there have not, as yet, been any studies of the discourses of LA21. The literature to date has concentrated almost entirely on documenting activity under this banner and analysing contrasts and comparisons (apart from Selman's work which will be discussed below).

The example that will be taken is from the English Lake District and, in particular, the South Lakeland Local Agenda 21 Strategy 2000, published (but not solely prepared) by South Lakeland District Council. Indeed, this document is particularly interesting because, while it is centred on the South Lakeland district and its district council, it is also clearly based in the involvement of a great many people and organizations in and around the area. There is a clear thread of communicative rationality running through it. The Foreword by a local academic states: 'Perhaps the most urgent requirement is for local communities to actually experience the basic principle behind Local Agenda 21, that it is those communities that should have the final say in what happens within their communities.' Similarly the Mission Statement—set out on the opening page—clearly states that 'South Lakeland District Council will consult and involve local people to help improve the well-being of all who live in South Lakeland whilst taking into account the capacity of the environment to support human activity both today and in the future'. Quotes on sustainable development from international organizations and the emblematic Chief Seathl are set alongside one from Mahatma Gandhi: 'We must *be* the change we wish to see in the world' (pp. 2–3, original stress). And, after discussing the various departments of the district council, there is the statement, 'There is so much expertise in the area, it makes sense to work together towards a common goal. This is the ethos of Local Agenda 21' (p. 24).

The structure of the report reinforces this. In sect. 2, a clear link is made between the idea of sustainability and 'working together'. Sustainability is a 'vision' that people share, it is something that 'we' achieve, and, echoing Gandhi, change begins with 'you and me'. This section poses the open-ended questions: What is a sustainable society? How are we going to achieve it?

Section 3 is given over to a very detailed account of all the local partnerships focused on some aspect of sustainability, and the different organizations involved. A mass of detail is presented over some fifteen pages, more than half the whole document. Clearly demonstrating the involvement of different organizations is an important task for the document. Some fifty-five partnerships are described, reinforcing the reference in the Foreword to over 200 individuals, companies, authorities, and groups being involved.

However, it would be wrong to rest the analysis at the point of demonstrating how the discourse of communicative rationality is drawn upon. For the LA21 documentation itself seeks the legitimation of other rationalities, particularly procedural rationality, and this relates to the institutional problems of putting LA21 into practice, organizationally and discursively. The influence of procedural rationality will be discussed first, followed by analysis of these institutional constraints.

Procedural Rationality and Local Agenda 21

The way in which procedural rationality is drawn upon along with communicative rationality is already apparent in the A21 quote given above. To restate it:

> 28.3 Each local authority should enter into a dialogue with its citizens, local organizations and private enterprises and 'adopt a Local Agenda 21'. Through consultation and consensus-building, local authorities would learn from citizens and from local, civic, community, business and industrial organizations and acquire the information needed for formulating the best strategies. The process of consultation would increase household awareness of sustainable development issues.

In addition to the points made above, it is clear that the full implications of communicative rationality, in terms of moving towards mutual understanding, are somewhat undermined by the use of terms such as 'consultation' and the suggestion that a key output will be information to support strategy development. These words are more resonant of procedural rationality.

Indeed the rest of the paragraph reinforces this point:

> [28.3] Local authority programmes, policies, laws and regulations to achieve Agenda 21 objectives would be assessed and modified, based on local programmes adopted. Strategies could also be used in supporting proposals for local, national, regional and international funding.

The earlier paragraph on the 'Basis for Action' emphasizes 'the participation and co-operation of local authorities' and points out:

> [28.1] Local authorities construct, operate and maintain economic, social and environmental infrastructure, oversee planning processes, establish local environmental policies and regulations, and assist in implementing national and sub-national

environmental policies . . . they play a vital role in educating, mobilising and responding to the public to promote sustainable development.

A clearer statement of procedural rationality as justifying the central role of a governmental body would be difficult to find.

This emphasis on procedural rationality can also be found elsewhere. For example, the five characteristics of LA21 defined by the International Centre for Local Environmental Initiatives (a key promoter of LA21 activity) in 1998 were:

- the integration of environmental, economic, and social issues;
- the integration of different interests;
- the connection to the precautionary principle;
- a global dimension, with the global impact of local action being identified;
- the sustainable management of resources.

While the integration of different interests is itself defined in terms of 'a culture of dialogue and participation', the framing of all five characteristics is essentially in procedural terms (LASALA 2001).

Similarly the three documents mentioned above all have a strong reliance on procedural rationality. The UNED Forum document on multi-stakeholder processes (MSPs) for sustainable development reviews the use of these processes in some detail and argues for a 'meta-communication' above the substantive communication within the MSPs. This would establish the rules for communication:

Several reasons make it advisable to create space for meta-communication in MSPs. Groups increase their effectiveness if they work on the basis of an agreed set of rules—hence they need to communicate about the way they communicate. Meta-communication also allows space for dealing with problems which arise when members feel that others are not playing by the rules. (Hemmati 2001: 95)

This meta-communication or 'communication about communication' is seen as dealing with some of the messy problems of human interaction. For communication and decision-making 'are not merely rational processes and should not be approached as such. People's feelings, attitudes, irrationalities in information processing, and so on, need to be taken into account and respected' (ibid.). However, the solution to these messy problems is a retreat into procedural rationality, to set the rules for how people should communicate.

The full rhetoric-line for the Environment Agency document *Consensus Building for Sustainable Development* (Fig. 9.1) makes clear the extent to which communicative rationality is combined with procedural rationality here also. The important question is whether the former is not in practice subsumed within the latter, whether procedural rationality is not the overall framing justification. Certainly there is a dominant emphasis on developing rules, models, and guidelines for best practice by the Agency. In simple quantitative terms, procedural rationality outweighs communicative rationality in this document. But

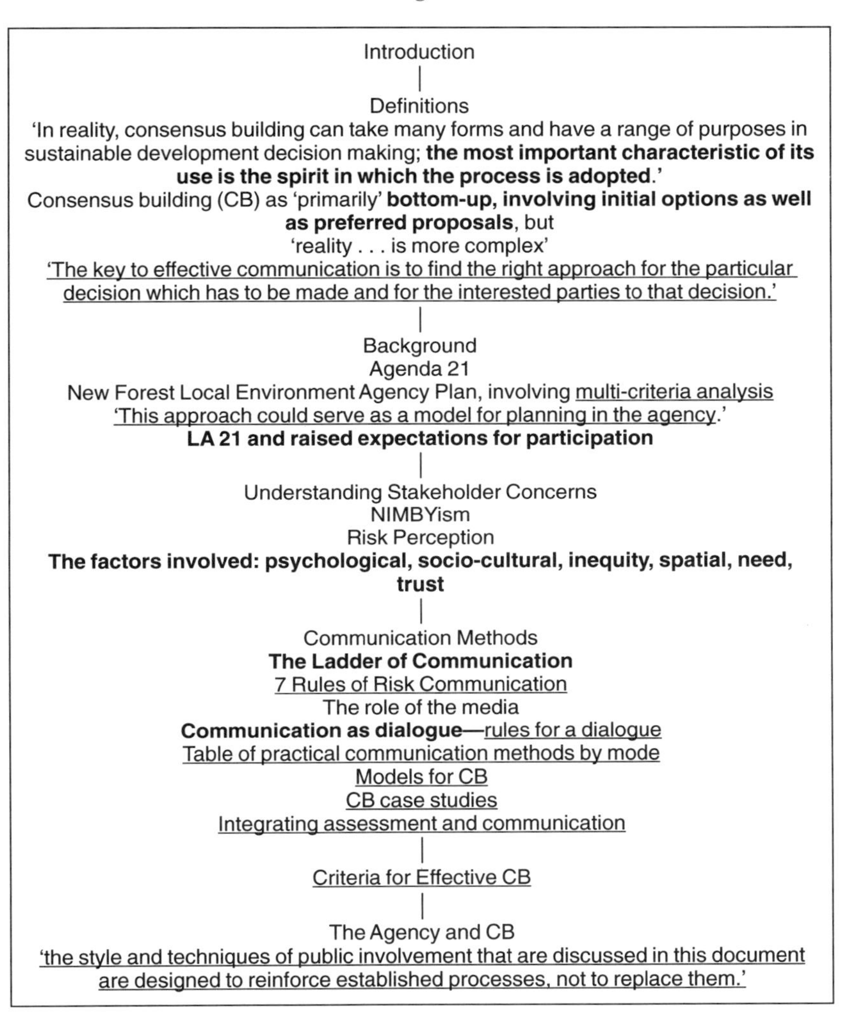

Fig. 9.1. A rhetoric line for *Consensus Building for Sustainable Development* (Environment Agency 1998)

Note: Underline indicates the rationality policy process discourse; **bold** indicates the communicative rationality discourse.

more significantly specific references to communicative concerns are immediately reframed in procedural terms: consensus building may theoretically be bottom-up but 'reality . . . is more complex'. Thus the message of Arnstein's ladder of communication (a much-cited 1969 article on public participation, which emphasizes moves towards community decision-making and empowerment) is followed by the Environment Agency's seven rules for communication.

The idea of communication as a set of dialogues becomes rules for dialogue. While not wishing to deny the Agency's commitment to trying to open up their processes of decision-making, it must be noted that the document conveys the argument: if this has to be done because of external pressure from the government and public expectations, then here is a procedure for handling stakeholder involvement. This is encapsulated in the last quote in Fig. 9.1. Procedural rationality clearly holds sway here.

Turning to the South Lakelands LA21 documentation, here the largest proportion of the document expounds communicative rationality (sects. 2 and 3 account for seventeen of the total twenty-eight pages), but procedural rationality is still at work. Section 1 frames the document by exploring LA21 and its origins, outlining the benefits of actually preparing a 'strategy' and discussing the procedural tasks of monitoring and reporting. This supports the work by Selman and Parker (1999), which used in-depth interviews with participants in LA21 processes to identify four dominant storylines. These focused on: participation and citizenship; new techniques for building participation and consensus; the pursuit of holistic policy solutions; and a reassertion of the role of local government. While the first two of these storylines draw on the communicative rationality discourse, the second two refer back to procedural rationality. The concern with holistic solutions (also noted in Bell 1998) relates to the ideal of the rational policy process as comprehensive and integrated. The reassertion of local government is a more overt statement of the need to retain control within the conventional boundaries of local politicians and policy officers.

In the South Lakelands document, because the two rationalities are generally divided into separate sections, it is not clear whether one is being given greater emphasis. But the concern of those committed to communicative rationality is that it may become subsumed within procedural rationality, that the latter may come to define the former. Bell sees this as a problem in the case of the East Midlands National Forest, where LA21 action plans, based on local focus groups, came into conflict with local government action to implement the National Forest Strategy. Similarly Patterson and Theobald (1995) point to the way in which the broader concern with procedural rationality within local governments can constrain innovative LA21 action. Their work considered the shifts in local government procedures and structures and its negative effects.

Any process or documentation that involves state actors will need to take account of procedural rationality, the dominant rationality by which the work of state policy actors is judged. This links into the way that the South Lakeland District Council activities are themselves presented in their LA21 document, in comparison to the presentation of the partnerships. Council activities are detailed department by department, under the headings of 'objectives' or 'best practice', with separate reference to the amount of cross-department work and integration across these departmental boundaries. All these aspects of procedural rationality relate to the underlying organizational belief that 'planning philosophy is central to achieving sustainable development' (p. 3). Thus, while communicative rationality is used to legitimate a wide range of activity across different actors, actual implementation is still based in planning and in the council, and is measured by the council's performance. The key point is that even if LA21 is a 'process not a plan', because local government actors are centrally involved in that process, LA21 will be subject to the same institutional pressures and, in particular, modes of performance assessment as other areas of local government activity. Therefore LA21 also has to be justified by reference to procedural rationality.

One significant dimension of this is the use of evaluation and monitoring tools within local government more generally and within sustainable development policy in particular. This is found in the South Lakeland's document in sections advocating the development and use of indicators to monitor and evaluate the council's performance in relation to the goal of sustainability. Reference is made to some thirty-five county council indicators and ninety district council indicators. Performance against targets for 1999/2000 is set out and targets for 2000/1 established. More generally, councils are making use of procedural tools such as sustainability audits and sustainability indicators to marry procedural rationality with a sustainability focus. Sustainability audit refers to an assessment of specific organization, programme, etc. from the point of view of its contribution (positive or negative) to sustainable development. Sustainability indicators refer to regularly measured aspects of sustainable development that can be used to monitor change and assess performance, and can be adapted for use as policy targets or evaluation criteria.

Such indicators are now found at almost all levels of government, from international (particularly European), through central, and down to local. The European Commission has several rafts of indicators, notably the Common Indicator Set for the urban level that they are currently promoting. The British Labour government have committed themselves to assessing movements towards *A Better Quality of Life* (the government's current Sustainable Development Strategy) by reference to some thirteen headline indicators, ranging from conventional GDP to wild bird populations and recycling rates. And many local governments see the use of locally developed indicators as a central element of their Local Agenda 21 strategies. The LASALA research found 63 per cent of their survey respondents (signatories of the European Sustainable Cities and Towns Campaign's Aalborg Charter) were using a system of indicators, mostly locally developed but also drawing on national government and European Commission indicator sets. These were being used for the purposes of inter-authority comparison, benchmarking against sustainability standards or goals, or measurement and monitoring of local progress (LASALA 2001: 57).

The general aim is that such indicators should be a new way of guiding or steering policy practice by reference to new goals. Policy practice will have to change since it will be evaluated and monitored by reference to these indicators. This could be a potentially radical development, bringing environmental goals into the heart of the policy process and increasing transparency and accountability by requiring policy actors to justify their actions with reference to these indicators. However, such indicators can be used in two rather different directions (Pastille Project 2001). One involves using innovative community participation techniques to bring new groups into the process of defining their view of sustainable development. Indicators based around themes such as 'quality of life' can become a means of community involvement, although the difficulties of collective action problems (Rydin and Sommer 1999) and contrasting oralities (van Woerkum 2002) always persist. The other is that they can become simply another set of indicators within prevailing New Public Management practices, which have introduced monitoring and evaluation by means of targets, criteria, and indicators across the whole range of public-sector activities (Stoker 1999). This means that sustainability indicators run the risk of being treated like all other NPM indicators and, as such, they merely become part of the paper-trail culture of government organizations rather than representing a radical shift in the prioritization of environmental policy goals. Similarly wider community involvement becomes rather attenuated, emphasizing accountability and transparency rather than any broader active involvement of local people.

Communicative Rationality, Local Agenda 21, and the Limits of Collective Action

However the limitations on the articulation and exercise of communicative rationality within LA21 are not just due to the presence of and reliance on procedural rationality. There are also the constraints arising from the institutional limits to collective action and the way that these can undermine legitimation in terms of communicative rationality.

According to communicative rationality, the criterion of a legitimate LA21 is the involvement of local communities. This can lead to a sense of failure if that participation is not forthcoming and this, in turn, can encourage a return to procedural rationality. Thus in the Environment Agency's document, it is stated that 'Experience indicates that the majority of public will take no interest in a decision-making process' (EA 1998: 3). The discussion then moves quickly on to ways in which the agency can learn to understand stakeholder concerns, in the absence of large-scale public involvement. Yet, given the nature of the collective action problem discussed in Chs. 4 and 5, participation is not always forthcoming. Sustainable development is both a vague concept and explicitly relates to issues of futurity. It also implies some knowledge of the ecological processes that may give rise to concerns about the sustainability of current practices. Lack of knowledge and the imbalance between the apparent benefits of participation and its costs have indeed constrained involvement in LA21. This has led those seeking to promote LA21 to emphasize its educational role, particularly where more captive audiences, such as schoolchildren are involved, and to promote activities that have a 'fun' or social dimension, such as Green Fairs (LASALA 2001: 37). It has also led to a recasting of sustainable development in terms of quality of life to render the concept both more meaningful to local communities and of more immediate concern. Selman and Parker (1997: 176) have emphasized the need to frame environmental issues in these terms in order to resonate with people's expressed preferences (see also Rydin and Sommer 1999 and the discussion on environmental capital and capacity in Ch. 8). However, even then, it has often not proved possible to generate substantial levels of public involvement.

A similar problem has arisen with the business sector. Most studies of LA21 practice have specifically identified the problems of involving this group. As with members of the public or communities, the patterns of incentives facing business often do not favour participation. Firms and business representatives always set the opportunity cost of time spent in LA21 activities against the likelihood of demonstrable sectional benefits, measuring both in terms of the contribution to profitability. The LASALA research project quotes a LA21 workshop participant as saying: 'We try to co-operate with the commercial sector but at the moment they mostly care about their business and profit' (2001: 29) and concludes that:

> By definition, 'stakeholders' will tend to involve themselves if 'stakes' are perceived to be important, and it can be assumed that interest organizations such as business and commerce are in the main little interested in LA21. This is because they cannot see its relevance to their activities and, with some exceptions, they are not prepared to invest resources in something that they view as peripheral. (p. 128)

Given the nature of many LA21 activities, both in process and output terms, the demonstrable, short-term, and financial benefits to business will be difficult to identify.

There is an additional discursive problem facing business, concerning the relationship between economic and communicative rationality. The essence of economic rationality is that the market is the best location for decision-making, both theoretically and practically. Even if certain environmental costs have to be internalized through taxation, for example, this does not remove decision-making from the marketplace, which remains in a privileged position. Such a situation works directly against the central tenets of communicative rationality, where all voices are on an equal footing and the goal (at least in the first instance) is mutual understanding. Why should an economic actor need to understand the position of other non-economic actors, given that it cannot impact on market-based decision-making?

This is not simply a case of actors favouring only strategic communication oriented towards supporting their interests. For community groups too have their own interests and we have seen how this can skew their use of communicative rationality, resulting in a position which may be termed NIMBY rather than fully based in communicative action. Nevertheless, for community groups who are seeking a voice from a position of relative silence, communicative rationality offers a justification for their involvement in policy debate. That justification has to sit alongside the other incentives facing communities, but it still has a role to play. But for economic actors, they have no need of communicative rationality to justify their involvement in policy. Economic rationality can suffice. This provides an underlying discursive basis for the constrained involvement of the business sector in LA21 activities. They simply do not need to adopt the performative attitude assumed within communicative action, because of the rival and preferred claims of economic rationality.

There is though another side to the problem of relying on communicative rationality. For such a reliance raises the question of whether LA21 should be judged primarily in terms of process rather than output. As organizations gain more experience with LA21, many have become impatient with the substantive impact of the considerable efforts going into the process. They want to see some results in terms of altered policy decisions, improved environmental quality, and changed behaviour in relation to the environment. This has also encouraged a shift towards procedural rather than communicative rationality and a desire to see LA21 feeding into mainstream governmental practices. This reveals a more general tension at the heart of communicative rationality between its support of deliberative and collaborative processes and its actual ambivalence in relation to substantive outcomes, whether social or environmental. Dobson (1998) makes the point clearly that a pursuit of process as a policy goal carries with it no necessary implications about outcomes. Indeed, this is a point that both Habermas and Dryzek, theorists of communicative rationality, accept, although seeking to suggest constitutional ways of remedying this (see the discussion in Ch. 2). Others, however, continue to cling to the hope that extending communicative rationality will imply progress on ecological goals also. Mason (1999: 9) strongly argues that environmental democracy 'extends and radicalizes existing liberal norms in order to include the ecological and social conditions for civic self-determination', and that this means more sustainable outcomes. This rests on the understanding of environmental interests as common interests (ibid. 48). However, his own study of LA21 in Islington, London, shows the considerable deficiencies of environmental democracy in practice.

The South Lakeland document also reveals this tension, lending weight to Dobson's view. For example, the Foreword is immediately followed by a 'message' from the Chief Executive of the district council. The Foreword states that 'It is vital that we heed the advice of scientists who point out that the current use of resources is dangerously close to the earth's biological limits.' It unequivocally argues that 'While we may be reluctant to admit that our economic and social aspirations are at the heart of the problem, time is running out.' Meanwhile, the Chief Executive takes a much more sanguine view in relation to environmental sustainability: 'Sustainable development requires us to find new ways of thinking and acting so we achieve social inclusion, economic survival whilst at the same time, protecting the environment.' This is also reflected in the Mission Statement, where environmental capacity has a qualifying rather than fundamental role. It is not at all clear that a more inclusive and communicative approach will result in more people agreeing with Professor Jeffers, who wrote the Foreword.

This discussion highlights yet again the problems of relying on communicative rationality in practice and hence the potential limits on deliberative democracy both as a form in itself and as a way of achieving sustainable development. It is, therefore, not entirely surprising that there is an increasing emphasis on procedural rationality framing arguments based on communicative rationality, rather than the other way round. The danger is that the reframing will be so extensive as to render the references to communicative rationality merely token. Communicative rationality will become a legitimization for prevailing or marginally altered practice, in the sense of cloaking what is actually going on, rather than justifying a new form of planning practice. Analysing the 'stories' of different participants in LA21, Selman and Parker (1999: 58) clearly also see this danger. They see the current balance in these terms: 'a neo-modernist desire to gain competence in emergent methods which, whilst requiring a degree of humility and vulnerability on the part of the officer, nevertheless recognise the crucial role of expertise and professional competence'. Perhaps this places too much hope in the idea of public-sector bureaucrats being willing to be humble and vulnerable. Rather more likely is the tendency that Burgess, Harrison, and Filius (1998: 1448) identify in their comparative Anglo-Dutch study: 'these case studies reveal how easy it is to assert the communicative styles of instrumental rationality [akin to procedural rationality as defined here] of most experts and professionals even in these innovative fora'. They quote one of the officials involved in the research as arguing 'you need a dictator . . . or everyone will find arguments to go on as usual' (ibid. 1456). This is hardly the message of communicative rationality.

Scientific Rationality and Local Agenda 21

One source of these calls for expertise and professionalism is the assumed ability of procedural rationality to deliver policy impacts, but another is the assumed connection between the limits of public participation within LA21 and lack of knowledge about sustainable development. This chapter, therefore, ends with a brief discussion of the role that scientific rationality can play in LA21.

There are two different calls on scientific rationality within LA21, each with its own dynamic. First, there is the sense that scientific rationality can provide information for those engaged in the policy process, primarily those within government structures or managing governance processes. Here the view of scientific expertise is as an input into policy, scientific rationality as framed within procedural rationality. The LASALA (2001: 33) research identified this as a key issue, both in terms of the lack of knowledge and expertise within local government and the need to work with outside agencies, particularly universities, to gain such expertise. But the second view is rather different and, to some extent, in opposition to the demands of procedural rationality. This sees scientific expertise as playing a role in motivating actors to engage in LA21 processes. Such knowledge can persuade key firms and organizations that they have an interest in being involved in local partnerships. For example, knowledge about a particular technology or method of waste management may encourage firms to become involved in a local waste project. The LA21 process is here reducing the transactions costs involved in acquiring such technological knowledge and this alters the incentive structure for involvement in the process.

And scientific expertise can be centrally involved in motivating more general public participation. Here though the emphasis is on conveying the sense of environmental and ecological damage that will result from unsustainable practices, and thereby motivating involvement and possibly behavioural change. Environmental NGOs have made considerable use of scientific knowledge in this way, marrying scientific to communicative rationality. In addition to the general authorization of future environmental scenarios through scientific discourse, they have developed a number of key concepts, which are based on scientific rationality for their legitimacy but are oriented towards mobilization. Ecological footprint, ecological rucksack, and environmental space are of this kind (Wackernagel and Rees 1996; McLaren, Bullock, and Yousuf 1998). Ecological footprint refers to the spatial area that is required to support the activities occurring within a city, region, or nation; ecological rucksack to the environmental burden that is represented by an individual's consumption; and environment space refers to the share of environmental resources and services that a city, etc., is entitled to based on an equal per capita distribution and subject to overall environmental sustainability constraints.

These concepts invoke scientific rationality to provide the knowledge base for identifying the link between environmental impact and the activities: all activities within a city, region, or nation in the case of ecological footprint; an individual's consumption basket in the case of ecological rucksack. On the basis of this rationality, they are presented as objective measurements and are intended to inform and mobilize. The aim is to highlight on a spatial or individual basis how much greater resource use is than might be expected. They all carry with them the implication of a preferred policy direction. They not only make a measurement but also provide the basis for a judgement. In the case of environmental space, an a priori assessment is made, according to scientific expertise, of the total amount of environmental resources and services that are available within the overarching constraint of environmental sustainability. This is then divided on a per capita basis, assuming equal entitlement to such resources and services. The resulting gap between actual use and this per capita entitlement is meant to inform policy. Environmental space, ecological footprint, and ecological rucksack thus all embody an assumption, informed by scientific rationality, concerning the limits to development and growth. Mittler states that 'to calculate environmental space targets, we must presuppose the scientific knowability of the limits to growth' (1999: 356; see also McLaren, Bullock, and Yousuf 1998: 12).

But there is more involved than just a call on scientific rationality. The purpose proceeding from these concepts is to steer the discourse towards an explicit acknowledgement of the interrelationship between ecosystems and societies across space: my consumption is your degradation (or vice versa). Implicit in this is the idea of the responsibility that one community owes another through the burden it places on the environment, and also the idea of some just or fair distribution of environmental benefits, services, and assets. Such concepts open up the discourse to voices beyond the local community and raise issues often not currently considered within the domain of local policy. This could be extremely challenging on both a conceptual level and in terms of the language used. Simmons and Chambers (1998: 355) note the 'richness of imagery' associated with these concepts and also the links with a broader romantic discourse.

An example of the application of one of these concepts can be found in the ecological footprint analysis undertaken for the Isle of Wight by a partnership comprising the Isle of Wight Council, Best Foot Forward (an NGO/consultancy) and Imperial College of Science, Technology and Medicine, funded by landfill tax credits through Biffaward. This exercise fits within the Isle of Wight Agenda 21 'Island Voices Speak Out for the Future', which is typical in its vocal commitment to communicative rationality and its attempts to marry this with a degree of procedural rationality (*The Agenda 21 Strategy for the Isle of Wight* n.d.). Determining and quantifying the island's ecological footprint involved considerable data collection, analysis within a model of resource flows, and final conversion into the footprint figures. From this Imperial College were involved in constructing alternative scenarios for the future. These had two purposes: first, to clarify in stark terms the extent to which current patterns on the island were unsustainable. And second, to suggest a raft of strategies for reducing the ecological footprint from almost two and a half times the sustainable average 'earthshare' down towards the global sustainable average.

While the report is largely an account of the technical exercise involved and the main quantifiable conclusions reached, the involvement of multiple actors is repeatedly stressed. For example, the scenarios and strategies developed by Imperial College 'illustrate the need for a collaborative approach between the many different actors involved in supply and consumption chains' (ibid. 48). They also stress the motivational role of the analysis: 'the calculations show that, even though individual measure may only have relatively small benefits, a combination of a range of measures could bring a significant reduction in the Island's overall Ecological Footprint' (ibid.). This is scientific rationality seeking to engage with the requirements and benefits of communicative rationality.

Yet concepts, such as ecological footprint, still run the same dangers as environmental assessment, capital, and capacity, that of incorporation into procedural rationality once admitted into routine policy practice. For all these concepts can be framed procedurally. In the case of ecological footprint, ecological rucksack, and environmental space, there is effectively an equation provided for turning the quantum of human activities associated with an individual, city, etc., into an aggregate figure representing environmental burden or entitlement. Simmons and Chambers (1998: 355) refer to 'a series of interacting mathematical algorithms capable of converting resources used to a land-area equivalent' when they describe their computer software 'EcoCal' for calculating ecological gardens, a variation on the environmental space concept. It becomes, therefore, a discursive struggle to maintain the radical potential of these concepts, with their associations with the global and with justice, while at the same time seeking to embed the concepts in planning practice and accept their legitimization in terms of procedural rationality. These issues are readdressed in the final chapter on developing a distinctive discourse of sustainable development.

The general issue involved here is the way that these concepts and procedures define the access of different groups to the policy process and the influence they have on planning practice. The question is how they are used in practice to balance the claims of communicative, scientific, economic, and procedural rationality and how they support or constrain the influence of different actors within the policy process. Such concepts do offer some prospect for shaping policy practice in new directions but they require active discursive maintenance to avoid being subsumed within procedural rationality or the prevailing politics of local interests. They raise the difficult problem of how to pursue goals close to the concerns of communicative rationality but within a context of procedural rationality, a problem already raised in Chs. 5 and 6. They also clarify the difficulty of achieving a complex goal, such as sustainable development, within the confines of procedural rationality. To date, most of the effort has gone into integrating community concerns and language under the umbrella of deliberative democracy, collaborative planning, or community empowerment. Perhaps the next round of discursive work needs to address how scientific knowledge can be opened up to debate within these concepts and procedures, allowing for new knowledge and ideas. This will be considered in the final chapter.

Conclusion

The Local Agenda 21 case study has shown the difficulties that are encountered when policy relies heavily on communicative rationality for its legitimacy. It has shown how this has become a strong theme in the justification of policy for sustainable development, and also how this faces problems of generating and sustaining collective action, particularly from economic interests. It has shown how the links that are made to procedural rationality can threaten to overcome the distinctive character of a policy approach based on communicative rationality, and it has highlighted that the nature of the calls on scientific rationality and the nature of the links to scientific expertise may be central in devising a distinctive discourse of sustainable development in which communicative rationality plays a part.

10

The Prospects for a Sustainable Development Rationality

The previous chapters have demonstrated the value of adopting an institutionalist perspective in which a role for discourse is fully recognized. While this is a broader claim that could be adopted in relation to any policy arena, this book has particularly sought to examine environmental planning. In this context, it has explored the ways in which policy is legitimated, the interaction of different interests, the incentives facing planners, the prospects for expanding public participation, and the constraints on more radical deliberative processes. Here, in this final chapter, a more normative turn is taken and the prospects for developing and embedding a sustainable development rationality are examined. Following the institutional discourse approach, this involves considering both the nature of a sustainable development discourse that could be used to legitimate policy effectively and the institutional arrangements for embedding such discourse in incentive structures that are relevant to key actors. The chapter ends by returning to the issue raised at the beginning of the book, concerning the role of environmental planning as a state activity within this perspective.

A Sustainable Development Discourse

There is a considerable literature that discusses the concept of sustainable development. Much of it tries to find the 'best' definition, which if operationalized as a policy concept would actually lead to the desired policy outcomes. For example, there is the familiar Venn Diagram definition, which sees sustainable development as occupying an area of overlap between economic, social, and environmental concerns. There is the Russian Doll definition, which sees environmental concerns as setting the context for sustainable development, surrounding the social and economic dimensions (Levitt 1998). There are models that emphasize the potentially necessary trade-offs between one dimension and another and where environmental protection is only achieved at the expense of economic development. And there are models that emphasize the potential for win–win scenarios, a new form of economic development that also delivers enhanced environmental protection (Gouldson and Murphy

1998). There are models that allow sustainable development to be locally defined, in terms of quality of life (Rydin and Sommer 1999), and those that see basic livelihood issues as central, particularly in developing countries (Hardoy, Milton, and Satterthwaite 1992). The social dimension can be defined strictly as material equity or extended to include rights to cultural autonomy. Sometimes the environmental dimension is seen in terms of the life support systems; sometimes this is extended to the support for meeting economic and social needs; and sometimes it extends even further into post-material concerns such as landscape appreciation (Owens 1994).

These all constitute attempts at a normative discussion of sustainable development, answering the question: How should sustainable development be defined? But, from the discourse perspective adopted in this book, the interesting question is: What is the rhetorically strong way of defining sustainable development, so that it can be used to support legitimation claims? How can a sustainable development rationality be generated, developed, and embedded so that it supports policy and planning practice? In the first chapter, the discourse of common environmental interests was examined and found to have significant rhetorical strengths. However, it has proved much more difficult to embed this discourse as a form of policy rationality. It was argued that this is partly because the rhetoric of common environmental interests works as a kind of reframing, an inherently limited discursive strategy that fails to engage with the nature of environmental conflicts and the ways in which they are socially constructed. And at the level of specific environmental policy issues, the strategy fails to take account of the institutionally defined incentive structures facing policy actors. Developing a rhetorically successful discourse of sustainable development involves much more than just being creative with words and skilled with rhetorical tropes. This is an important element and will be dealt with shortly. However, developing a successful discourse also involves understanding the institutional context in which those words and tropes will be used. These issues will be returned to later in the chapter.

The immediate task is to consider the rhetorical strength of a sustainable development rationality. This cannot, of course, be conducted in complete isolation from the actual content of that rationality. After all, if one seeks to persuade, the way in which one persuades (the rhetoric) has to be related to the subject of persuasion (the content of the argument) as well as the object (the audience). How is this subject of a sustainable development rationality to be constituted? The distinctive content of sustainable development as a concept lies in its holistic nature, the way in which it seeks to combine the environmental, the economic, and the social, as has been touched on above with reference to the Venn Diagram and Russian Doll models. These three dimensions—the environmental, the economic, and the social—are closely related to the three rationalities that have been examined in terms of their individual legitimation of environmental planning. Scientific rationality supports the claims of environmental sustainability; economic rationality relates directly to the economic dimension; and communicative rationality justifies the involvement of a broad range of actors and consideration of a wide range of perspectives, a key link to social sustainability.

This enables the question of the content of a sustainable development rationality to be recast in terms of how these three substantive rationalities can be combined, discursively, to produce a new distinctive rationality. This is not the same task as deciding whether the different rationalities can be combined logically, in terms of their content or assumptions. While this is clearly relevant, it would be a work of normative theorizing and is not the prime concern here. Rather this section considers how the discursive structure of these rationalities affects their potential for being combined and, therefore, used in discursive strategies of rationalization within environmental planning. It is an inherent characteristic of the discursive domain that multiple and innovative ways of legitimating policy will be tried out by actors. Some of these will fail and some succeed. Such creative use of policy discourses by actors is one reason why any attempt to map the terrain of environmental policy discourse will produce a number of interrelated and even overlapping discourses (Myerson and Rydin 1996*a*; Dryzek 1997). How can this feature be used towards the goal of sustainable development?

The aim is to build on the individual rationalities, which have already been used within environmental planning situations, to generate a rhetorically consistent new discourse. It is not a task of devising a reframing discourse but rather using the discursive dimensions of existing institutions of environmental planning. To do this requires a little more consideration of how discourses engage with each other and can be combined in new ways. One way of conceptualizing the interaction of discourses is as a process of challenge through critique and counter-critique, a dialectical process that can lead to potential new resolutions and positions that may have greater strength within the policy process. Thus the critical engagement of different institutionalized perspectives on the environmental policy can be part of an ongoing process by which a stronger justification for that policy can be built. This idea can be explored by considering the way in which the three rationalities already discussed could and already do engage with each other rhetorically.

Starting from Scientific and Economic Rationalities: The Sustainable Development Rationality of Affluence

The discussion begins with the engagement between scientific and economic rationalities. This has been a very dynamic site of interaction within environmental debates, and we can clearly discern position and counter-position, critique and reformulation. The critique has taken the form of a far-reaching interrogation of the economic paradigm from the perspective of environmental and ecological science. It suggests that economic rationality, while correct to focus on the significance of economic processes, fails to recognize the fundamental constraints placed on such processes by the natural, physical, material world. The role of scientific rationality lies in identifying the limitations that exist to the uninterrupted continuation of economic processes, as envisaged by economic rationality. These limitations are linked to the physical nature of the environment and of ecological processes, as revealed by scientific knowledge, encompassing the irreversible nature of much damage to the environment, the danger of massive levels of change occurring within ecological systems as a result of the aggregate effects of many other small changes, and the catastrophic effects of passing through thresholds in terms of ecological change and damage.

The critique draws on a distinctive environmental discourse, the 'limits to growth' discourse, named after the *Limits to Growth* report (Meadows *et al.* 1972) and has been analysed by Killingsworth and Palmer (1992: ch. 7), Hajer (1995: 80–4), and Dryzek (1997: ch. 2) as a discourse with considerable rhetorical strengths. It is linked to a developed theoretical debate, which engages with and challenges the theoretical neoclassical basis of economic rationality as in the work of Daly (1977) and Martinez-Allier (1987). Daly's model explicitly seeks to integrate the laws of thermodynamics, including the law of entropy, into an understanding of economic processes; it, therefore, marries scientific and economic rationality. The reason for modifying the standard economic model derives from the knowledge of the physical world provided by science. Hence, the 'limits' discourse has repeated recourse to scientific metaphors.

However, this is a discourse intended for a broad audience, including policymakers and economists and, therefore, this is not a highly technical discourse. The scientific metaphors are populist, not technical in nature. Scientific knowledge is used as a 'taken for granted' accredited basis for a revised economic model. Similarly, however erudite such work is, it is notable that it avoids the abstract theorizing of the neoclassical school even when explicitly making linkages to that school's models. Rather it adopts a more approachable style, linking populist appeals with emotive language, economic argumentation with ethical claims. The authority of this discourse lies in its breadth, the bringing together of parts from different disciplines and approaches, and its explicit espousal of the holistic approach. This means that synecdochic rhetoric is a significant feature with relatively brief references to an issue being used to support discussion on a wider front. Thus reference to the laws of thermodynamics is used to stand for all scientific knowledge and to support a new economic model. This form of argumentation can leave a sense of skimming the surface of issues and not developing points in depth. Indeed Killingsworth and Palmer (1992: 243) argue that Daly's work 'has the feel of a utopian text'. However, this need not be a problem; many highly successful populist discourses take this form.

There is an alternative engagement between economic and scientific rationalities, which starts from the perspective of economic rather than scientific rationality. It is a more optimistic discourse, which seeks to incorporate the knowledge generated by environmental science within the prevailing dominant economic models. Many of the points made by the 'limits' school are, in fact, also made by environmental economists such as Pearce, who goes on to argue that valuation techniques enable the appreciation of such limits to be fully integrated into decision-making, via market-based policy instruments. We have also seen a more practical version of this line of thinking within pollution control (Ch. 7) where economic cost is seen as a constraint on applying technological solutions to unacceptable emission levels. Here a practical form of scientific rationality is used to solve an environmental problem, but only within the limitations set by economic rationality.

As with the 'limits' discourse, there is little rhetorical difficulty in connecting the two rationalities in this way. As at a theoretical level, the two rationalities can complement each other because they have a similar rhetorical form. They both adopt the ethos of the expert, so that the combined rationality can also speak with the expert voice. They both adopt synecdoche with realist assumptions, so that either scientific or economic expertise reveals the processes actually at work in the natural or economic world. And since this expertise operates in different domains, they can supplement each without disrupting these assumptions. And the form of closure used in arguments is also very similar: expertise provides knowledge of the problem and of the solutions. In the more practical version, again, there is potential for a complementary rhetorical relationship between the two rationalities. Practical experience identifies the potential for finding solutions that are both technologically feasible and economically viable in real-world situations. Discursively economic and scientific claims to legitimate policy approaches can live together.

Discursive compatibility has supported the development of a recent resolution to the choice between a 'limits' discourse and the more optimistic environmental economics version: the ecological modernization discourse (see also Hajer 1995; Jänicke 1996; Dryzek 1997). The essence of the ecological modernization storyline, as Hajer terms it, is that technological change can radically increase resource efficiency, leading to a new path of economic and industrial development. This would take the form of a restructured green economy, investing in new energy sources, producing goods and services for 'green' demand from consumers and producers, and comprising clean, efficient businesses. It is, in effect, a more dynamic version of neoclassical environmental economics, one that sees the potential for economic and technological restructuring in order to avoid the constraints of ecological limits. It is an appealing prospect, which is, not surprisingly, very popular with businesses. It is a highly optimistic discourse that emphasizes its practicality and applicability, for example in the 'Factor 4' publications of von Weisacker, Lovins, and Lovins (1998). At the same time, ecological modernization stresses its basis in the twin knowledges of economics and science.

This does, however, raise difficulties when seeking to engage with communicative rationality also. As has been discussed above and will be highlighted again below, communicative rationality challenges claims to expertise, seeking to open up decision-making to the broader involvement of many social groups. Indeed it seeks to recast the policy process as something rather different, deliberation rather than instrumental decision-making. This sits uneasily with the essentially instrumental language of economic and scientific rationality; similarly with the rights-based language at the core of communicative rationality. However, it means that the sustainable development rationality deriving primarily from an engagement of scientific and economic rationalities has only a subsidiary role for communicative rationality. Social concerns become defined through expertise, usually economic expertise as in valuation exercises. Or the 'social question' becomes one of communicating with social groups, as in the traditional engineering model of risk communication. The challenge to the ecological modernization vision that might arise from adversely affected groups or those with a different vision is silenced. Hence, this particular optimistic development of a sustainable development rationality can be termed a discourse of affluence.

Starting from Communicative and Economic Rationalities: The Sustainable Development Rationality of the Poor

A very different discourse arises if one begins from the interaction between communicative and economic rationality. The critique of economic rationality from the perspective of communicative rationality assesses the extent to which different groups and actors have been sufficiently taken into account in economic decision-making. Because communicative rationality is essentially concerned with how decisions are taken rather than just what their outcomes are, this critique does not simply take the form of assessing inequalities in those outcomes. This is often an element of the critique but not the core element. If outcomes were the key focus then inequality in those outcomes could be addressed through some form of procedural rationality alone, such as using indicators to monitor outcomes. However, communicative rationality emphasizes the importance of the range of stakeholders being involved in the decision-making process and this poses a significant challenge to the autonomy of market decision-makers, an autonomy that is at the core of economic rationality.

Within the environmental policy domain, the environmental justice movement provides a good example of this critique of economic rationality from the communicative perspective. While most accounts of this movement trace its origins back to the mobilization in the USA of people of colour against the location of toxic waste sites near their home (e.g. Harvey 1996), environmental justice can also refer to many other situations:

- the more general issue of the location of environmental bads *vis-à-vis* different social groups;
- the campaigns by local communities in developing countries against major development projects that will displace them; and
- the analysis on a global level of the adverse impacts of international capitalism (usually discussed under the heading of globalization).

In each case, the identification of unequal outcomes is married with a challenge to the way that economic decisions are taken and a call for wider participation in those decisions by all affected communities.

While these calls embody the claims of ecological or discursive democracy, they also represent a developed critique of contemporary economic processes. As such they suggest the hope that an alternative path of economic development, one that is truly sustainable from a social perspective, is possible. This is a radical critique of many existing models of development, including ecological modernization and even the Brundtland Report's calls for a 'new era of economic growth' (De La Court 1990). Instead environmental justice on a local and global level calls for a serious consideration of truly alternative paths of development and a questioning of prevailing processes of exchange and international trade.

This is a highly politicized form of argumentation, based in the concept of constitutional rights (derived from communicative rationality) but heightened by the awareness of inequality resulting from current economic processes. This points to one of the rhetorical weaknesses of such combined discourse. The discourse sets up an opposition between the political and economic status quo on the one hand, and marginalized groups on the other. This can readily be dissolved into a more overt ideological confrontation, between right and left, establishment and counter-culture, conservative and radical. The restatement of the debate on economic rationality in such simplified forms can then short-circuit the impact of the challenge. The critique in terms of environmental justice becomes opposed by a position (based in economic rationality) which argues that these groups are simply pursuing their own interests at the expense of a broader public good (represented by economic development). Argument and counter-argument become transformed—by association—into a familiar and irresolvable dispute between established ideological camps.

An alternative approach is to focus on the positive project of reformulating economic rationality from the perspective of communicative rationality. Such an approach is based on a theoretical critique that rejects the neoclassical model with its division of economic from social processes and argues that its view of actors' decision-making is fundamentally flawed. Market-based policy instruments and valuation techniques, such as contingent valuation methodology, are based on the idea of the individual household or firm rationally comparing costs and benefits at the margin. The critique of this theoretical and methodological framework is grounded on the argument that values are not sourced in individuals' decisions but are socially constructed. The emphasis is on the processes of social interaction that lead to particular values emerging, being defined, and prevailing (O'Neill 1993; Foster 1997). This suggests that a methodology that replicates these social interactions will be a better way of gaining access to these values; currently, the focus group or a variant is the preferred methodology. Thereby, the emphasis is also shifted away from the monetization of environmental policy, represented by economic rationality, towards a broader basis for establishing environmental values. The critique of economic rationality from a communicative perspective parallels that of scientific rationality reviewed above. Can this be as significant a challenge to economic expertise as it has been to scientific rationality?

To be rhetorically effective, the communicative challenge to economic rationality would need to find a way of re-expressing economic expertise that meshes with the rights of people to become involved in decision-making that affects them, their livelihood, and their quality of life. This could be achieved through a re-emphasis on the institution of the community. Such communities would be the principal sites where valuations are created and rights are exercised, in line with communicative rationality. But further, it would involve seeing the community as the site where economic processes are experienced and also constituted; that is, economic processes only occur because of the social interactions between actors within the community or communities. Such communities can be variously defined, of course, as communities of place, of identity, or of interests (Duane 1997). Whatever the specific focus of any discussion, the rhetoric would involve a reorientation from the level stressed in neoclassical economics of the market or economy, down to the community.

This would be a discourse of community economics, rights, and values. The community's enhanced role in environmental planning is justified by reference to its experience of the environment but also its constitutional rights to that environment, which are prior to other rights over the environment. This means that a constitutional language of rights and of common subjects ('we' not 'I') has to structure the discourse. And the community also becomes the basis for expressing the nature of economic processes, so that the more abstract (and abstracted) language of conventional economic rationality is replaced by language more directly based in experience of production, exchange, and consumption. Elements of such a discourse are already found within the considerable (but largely marginalized) 'small is beautiful' environmental literature (Schumacher 1974; Ekins 1986).

Such a discourse is, as might be expected, a significant challenge to economic rationality and is likely to face considerable resistance. There are a number of reasons for this. First, there are the linkages between the expert and everyday economic discourses noted in Ch. 6, which reinforce each other. The everyday economic discourse is the way in which most people currently interface with economic rationality. It can thus act as a buffer for more general criticisms of economic rationality. If the discourse of corporate discretion appears plausible, it becomes more difficult to challenge the expert discourse. Second, economic expertise has found itself more sheltered from criticism than scientific rationality, which has found itself held directly accountable for uncertainties in the application of scientific knowledge. Because economic management is a more directly politicized arena than scientific management, governments, rather than the proponents of economic rationality, are held to account for economic failures. Scientific policy failures are still traced back to scientific expertise and to the failure to use it in the right way; economic policy failures are seen as internalized failures of governments. This, in effect, insulates the theoretical discourse of economics from political as well as social critique. This does not mean that economic rationality cannot be challenged within the policy agenda. It just means that it will be more difficult to maintain a successful comprehensive challenge, as implied by a discourse of community economics, rights, and values.

For this reason—and as befits its localist message—this alternative discourse is mainly being pursued in many different locales. It is becoming real, not through a general discursive campaign, but by becoming embedded in many local institutions. For example, it is becoming a medium of communication in many initiatives involving community ownership of resources, community-based arrangements for decision-making on production and resource use, and community-level alternatives to exchange through market processes. There is now a range of such initiatives reported in the literature (Ostrom 1992; Young 1997). The common thread to these is a critique of mainstream economics combined with a commitment to communicative rationality, and an attempt to mesh deliberative forms of decision-making into local economic processes of consumption and production. The grassroots perspective of this discourse makes it pre-eminently a rationality of sustainable development that is sought by and resonates with marginalized groups in society. This is a rationality of the poor, seeking to legitimate policy action in the name of sustainable development from the perspective of those currently disadvantaged and without power.

What is the potential involvement of scientific rationality in this discourse? To engage with the sustainable development rationality of the poor, the claims of scientific rationality would need to be expressed with some humility, so that the ethos of the all-knowing scientific expert could be dispensed with. Rather, a place has to be found for community-based knowledge about the environment, alongside conventional scientific expertise. The direct engagement of the individual and the community with the environment is seen as the basis for wisdom arising from lay sources, so that wisdom comes from experience of the environment, not just the study of it. As we shall see, some similar points arise in the final discourse to be considered.

Starting from Scientific and Communicative Rationalities: The Rationality of Ecological Safety

It has already been mentioned that scientific rationality finds itself both a key plank of legitimization of environmental policy and being challenged for its justification of a dominant position for scientific élites. These challenges arise in large part from the failure of scientific institutions to live up to the expectations they have aroused. Lived experience of the gap between the image of scientists as objective, knowledgeable, and capable of right judgement and the way that various events have played out undermined this image. Nuclear accidents, food scares, environmental disasters, and medical errors have all featured doubts about the judgements, knowledge, and practice of scientific experts. Local communities have challenged the way in which scientific expertise has been used to shape the planning of their locales. Environmental NGOs have been significant in highlighting these doubts. They have also highlighted the fact that scientific knowledge is not unitary, that there are conflicting views and opinions within the scientific community. And while NGOs have often used their own scientific experts and claimed that they have the better scientific argument, this has proved to be a double-edged sword. The NGOs may be trying to use the claims of scientific rationality but they are also demonstrating its contingent nature; the claims are different in substance if one listens to a Greenpeace scientist or a governmental scientist.

In so far as the challenge to dominant scientific expertise, say within governmental or commercial organizations, comes from scientific expertise in NGOs, it is a clash between different versions of scientific rationality. The discourse of that rationality remains largely unaltered as, say, Greenpeace dispute with Monsanto about the current state of research and knowledge on GMOs. Wherever it is based, scientific rationality continues to envisage a knowledge gap between expert and lay groups. The critique of communicative rationality by scientific rationality remains one couched in terms of lay ignorance, with scientific rationality being the route to filling the knowledge deficit. However, communicative rationality offers a challenge to the idea that knowledge is based in scientific rationality alone. Instead it offers a view of knowledge as plural, not in terms of different scientists engaging with each other on the basis of scientific rationality but in the sense of non-scientific voices having a legitimate role, as discussed above.

Such a position can lead to the very notion of knowledge becoming plural and diverse, rather than ultimately unified, so that a number of different communities may develop their own criteria for knowledge generation. In this case, it is no longer possible to choose between these knowledges according to overriding and universal criteria. The emphasis then falls on open and inclusive communicative processes that cover the full range of knowledge claims, including 'lay' knowledge. These open and inclusive processes then have to adjudicate on the appropriate balance between different knowledges, but where such a balance should lie is an open question subject to communicative agreement in each case. Here, communicative rationality is acting as a displacing, bottom-up alternative to any top-down discourse, including the discourse of specialist scientific expertise. This is a potentially radical challenge to scientific rationality's claims to knowledge.

There is a resolution possible, though, between legitimating policy in terms of scientific expertise on the one hand, and the engagement of plural voices on the other. These involve claims for greater lay involvement in scientific debates alongside and continuously engaging with scientific expertise. This mutual engagement would also extend to issues of how scientific expertise is used within the policy process. Such a revised institutional basis for involving science within the policy process has been termed 'citizen science' (Irwin 1995). This 'opening up'—but not replacing—of science means that communicative processes are used to interrogate the quality of the science and to distinguish good from bad science by assessing the institutional context within which scientific practices take place. It also supplements the inevitable limitations of scientific expertise with the knowledge gained from local lived experience, so that scientific knowledge is enhanced and the total knowledge base broadened. The status of 'good' science remains high and it is accorded appropriate weight within policy deliberations but becomes subject to double legitimation by its own ideal of knowledge creation and the claims of communicative rationality.

Such a resolution may not be easy to sustain rhetorically. The unqualified commitment to participation in policy debates is presented in communicative rationality discourses as a constitutional right. It therefore sits uneasily with the specific claims to knowledge arising from scientific rationality. It may be possible to relegate these to different realms—science providing information, participation providing judgement—but this is to reduce the potentially creative engagement between the two substantive rationalities to one of procedure. It also downgrades some of the central elements of communicative rationality about the significance of different viewpoints really engaging with each other. The resolution requires a discursive solution that goes beyond the removal of scientific jargon in policy debates. It requires a form of legitimation that recognizes the particular contribution of scientific rationality in supporting knowledge claims but also the constitutional right for others to put those claims to the test. Knowledge creation—in a policy context—becomes a joint process of multiple actors and is legitimated in multiple settings. Claims by each side have to be recognized and a discourse of some considerable humility has to prevail as each side recognizes these claims and the rights of others to challenge them.

In effect, this involves providing a constitutional language for recognizing the rights of multiple parties to discuss knowledge claims and for deciding on those knowledge claims in any specific circumstance. Developing such a language has a wider significance than setting up organizations for debating the scientific dimension of any specific issue. For example, the various public and private inquiries into BSE and the foot-and-mouth crisis affecting British agriculture and the food industry may be seen as an organizational way of opening up discussion about the scientific issues involved. More important than such organizational arrangements is the creation of an institutional basis for 'citizen science' and this, as implied by the very concept of institution, requires the development of appropriate norms, values, and culture; that is, it requires a constitutional language of open science.

It is highly questionable how economic rationality can engage with this combined discourse. In many ways the concerns of economic rationality lie outside the discursive approach involved here. Indeed, it can be argued that this discursive position is not really a sustainable development discourse at all, because it does not really engage with economic issues, it talks past them. There is no linguistic point at which an engagement can readily occur. Sustainable development, as emphasized above, is a holistic concept that encompasses the environmental, the scientific, and the economic. A discourse that draws primarily on the environmental and communicative in order to justify positions and decisions cannot be considered a sustainable development rationality. Hence, it is more appropriate to label this a discourse of ecological safety, primarily protecting social and environmental concerns. This reliance on a more limited discourse will itself limit the outcomes that can be expected from policy debate. One of the problems that the environmental justice movement has had is that it has used both a sustainable development rationality of the poor and this more specific discourse of ecological safety. This may lead to some policy impact in terms of social and environmental protection but it limits the shift towards sustainable development.

Building Institutions for Sustainable Development

The three discourses provide rhetorically strong ways of seeking to justify policy towards sustainable development. There are clearly value judgements involved in choosing between them. But it is not enough just to identify a discourse and recommend its use within policy debates. Institutional analysis shows that, to be effective, a discourse has to be embedded in organizational arrangements giving voice to appropriate norms, values, and routines, and shaping practice. Thus for a new way of legitimating environmental planning to achieve any resonance, institutional arrangements will have to be built around it.

Much of this institution building will be highly specific to the particular issue at hand. Incentive structures facing actors are always context-specific. However, to approach the general question of what these institutions should be like, it is helpful to distinguish two aspects that are sometime conflated. On the one hand, there are the institutional arrangements that influence our use of the environment. These encompass the organizational arrangements, the sets of incentives, and the prevailing norms and values that explain how specific aspects of the environment are overused (from an environmental sustainability perspective) or not. We live with multiple sets of such institutions that affect the many facets of our use of the environment, from energy use, to waste generation, to holidaying in certain landscapes. In many cases, these combinations of institutional arrangements do result in free-riding and environmental degradation, and the challenge to environmental planning is to build new institutional arrangements that prevent degradation. The same point can be made about numerous individual actions that contribute (or not) to the goal of sustainable development. More generally, we need institutional arrangements that promote sustainable development, supported by a discursive justification.

This argument for environmental planning to build appropriate institutions for environmental protection or sustainable development has to be met by the appropriate supply of planning activity. But this is not necessarily straightforward. Throughout the book, the institutional arrangements for environmental planning have been discussed. It has been shown that, again, organizational arrangements, incentive structures, and prevailing norms and values influence planning practice. Particularly relevant aspects of this are the incentives for collective action in support of environmental planning within society and the institutional influences on planning bureaucrats. Designing environmental planning to promote sustainable development involves incentivizing and justifying state action, within planning bureaucracies and within society in general.

The key challenge is to link together the institutions of environmental planning, which influence the actions of key policy actors, with the institutions of environmental use, which influence the actions of those who use the environment. If these two sets of institutions are in accord with each other, then there is the prospect of movement towards key planning goals, such as sustainable development. If they are not, then planning activity may grow but not achieve policy goals or planning activity may be forestalled. This is a detailed task of institutional design for each aspect of environmental use and sustainable development. It requires close attention to the ways in which sustainable development may be promoted, to the incentives needed to achieve this, and the specific discursive justification used to support policy action.

The work of Elinor Ostrom has focused on the twin task of designing institutions of environmental use and environmental planning. But by concentrating on the community level, her otherwise very valuable work has created the danger of conflating the essentially distinctive character of these two institutions. Only with community-level co-operative action are the structures for collective action to protect the environment and the structures for use of the environment one and the same. Social capital can be developed in these circumstances that generates both local community planning for sustainable resource use and, at the same time, sustainable resource use itself. The state is effectively bypassed or only considered as an enabling state, encouraging this doubly beneficial situation to arise. But in most modern contexts the state and state planning for environmental protection or sustainable development are still necessary.

The book ends by considering the role that state planning plays within this perspective. There are a number of reasons why sustainable development will continue to need to engage with state planning. First, initiatives outside the state and at the level of the community will only be able to deal with certain aspects of sustainable development and will leave untouched aspects that transcend the community level. Many environmental concerns, in particular, cross boundaries between geographically based communities and raise issues that need to be dealt with at a regional, national, or even international level. Therefore, state organizations at these levels are the most appropriate focus of policy practice.

Second, the state is undeniably a powerful actor. But it is not the only powerful actor and many analyses of environmental policy (such as Hajer's, for example) can point to the power of the business and finance sectors and other institutionalized interests, such as the professions and scientific expertise. As Galbraith (1952) has argued, powerful actors should face equally powerful actors across the policy table if their power is to be kept in check. This idea of countervailing powers suggests that the state is necessary to constrain the influence of the already powerful and, thereby, to protect the interests of the less powerful within society. The fact that we have so much evidence of the state failing to act in such a way should not preclude the possibility that a reformed state could fulfil this role more competently.

Third, there is evidence, looking at states across the world, that some are able to incorporate a greater level of environmental concern than we might expect from our British perspective and are even able to take quite a radical perspective. Environmentalists might look with envy at the policy statements and actions of, say, some Scandinavian countries but this can be translated into hope, once one asks the question of how a sustainable development perspective can be fostered within the state.

Finally, it has become academically fashionable to point out that the state often lacks the 'capacity to act' (C. Stone 1989). But it should not be forgotten that, if one wishes to achieve particular outcomes, it would often be necessary to harness the power and capacities of the state to achieve these outcomes. This is not to say that there is an easy and ready transition from policy goals to outcomes, but rather that the state remains a central vehicle for seeking to achieve policy goals, however imperfectly. It is notable that many of the countries that are considered leaders rather than laggards in the environmental domain are those with both a great faith in state policy processes and relatively strong and effective state management mechanisms.

So there remains a need to consider how sustainable development concerns can be embedded within the planning processes of the state. This involves considering new arrangements for the involvement of actors in policy processes and new roles for environmental bureaucrats. It also involves new 'rules-in-use', both soft and hard, governing environmental use. Following the discussion above, two central aspects will be how to handle expertise (particularly scientific and economic expertise) and how to embed rights (particularly community rights) in institutions. Indeed embedding rights can shape the handling of expertise. Rights—and their corollary, duties—involve all three aspects of institutions: organization arrangements, incentive structures, and discourses of values and norms. Rights and duties perform a number of functions. They are a means of defining norms and acceptable behaviour. But they are also a way of defining relationships between actors. Any right is exercised by one actor *vis-à-vis* another; and any right defines a commensurate duty. At the same time, rights and duties are enforceable by reference to a higher or external authority, often but not necessarily the state. Rights and duties, therefore, encompass the moral, the legitimate and, within these constraints, the possible limits of behaviour. By creating, granting, and maintaining rights and duties, the state (or other authorizing body) has the potential to reshape relations between actors and embed those new relationships in institutions. Thus networks of rights can perform the embedding function in relation to a new discourse, particularly where that discourse itself centres on issues of rights and duties. The question is, what are the rights and duties that should be assigned to actors in order to support a policy justification couched in the particular discourse of sustainable development chosen?

The prospects of embedding the proposed new rationalities outlined above will depend on the creation of new institutions of rights. Promulgation of the new rationalities will help support these institutions and, dialectically, these institutions will maintain and reproduce the rationalities. The institutions require a discursive and incentive-based reality to survive. For the sustainable development rationality of affluence, these rights will be primarily defined in terms of being market-based, particularly as they relate to technology and the application of scientific knowledge. For the sustainable development rationality of the poor, these rights will be (as emphasized above) based at the community level. While for the discourse of ecological safety, it will be the definition of environmental rights, perhaps held by those designated as speaking for the environment, that will be most significant.

It is not enough, though, to consider only substantive rationalities. Where state planning is involved, there is bound to be an engagement also with procedural rationality. As has been shown at many points in the book, procedural rationality in the form of the rational policy process discourse has considerable influence on policy processes and outcomes. It can subvert other rationalities within the context of state planning. It is, therefore, significant to ask whether the rationalities discussed above run similar dangers of being reframed by procedural rationality. Again, it can be suggested that a focus on rights and duties may hold more scope for containing the discursive powers of procedural rationality. If the rationality of the policy process is judged in relation to questions of rights and duties—citizens' rights, rights of scrutiny or appeal, and rights to aspects of the environment—then this involves a redefinition of procedural rationality itself. Procedure is seen more in relation to the occasional yet specified involvement of external groups asserting rights, than to the routine tasks of public-sector bureaucrats. Such a redefined rationality would pose less of a risk of subsuming communicative rationality, when they came into contact. It would also be less able to combine readily with expert-led discourses, such as scientific or economic rationality against non-expert groups. There may be more scope for the claims of scientific and economic rationality to be considered alongside others during these procedures of scrutiny and appeal.

Such an approach may also help place the pursuit of participatory processes and, in particular, the attempt to build consensus in perspective. Such a pursuit can become a rationalization of policy practice, not in the sense of a reasoned justification but as a false representation of actual practice. There is a need here to maintain a critical stance on actual planning practice, while maintaining the commitment to more participatory forms of that practice; to be critical without being cynical; to have an ideal without being naïvely idealistic. One conclusion of the analysis of this book is that deliberative and collaborative forms of public involvement may not be the most appropriate in all but selected cases. Instead, it may be that public involvement should take an alternative form. Rather than work towards consensus—a consensus that is not achievable—it may be more appropriate to keep open spaces of contestation within environmental planning. Of course, too much contestation is problematic. It is expensive of time and resources, debilitating for participants, and gets in the way of action on policy issues. So, endless conflict and argument is not wanted. There is a real pressure for closure at points in the policy process. Agreements need to be reached and decisions made. However, it will probably be false to present these agreements and decisions as consensual, whatever the participatory arrangements involved. What is needed is a recognition that the agreements reached within the policy process are provisional and should be open to periodic debate and scrutiny. This involves establishing rights to scrutiny and periodic examination on the part of groups within society. Third-party rights for environmental groups to challenge environmental policy would be a key example here.

This is not to suggest that all hopes of deliberation should be jettisoned in place of a reliance on courtroom contestation on the basis of rights and duties. Where deliberation and collaboration can realistically be made to work over specific environmental issues, they should be encouraged. Furthermore, there may be one area where deliberation, in particular, has an important role to play. It is part of the argument for public participation and deliberation that actors' values should be accorded greater significance within the policy process. From this follows the recommendation of focus groups and other techniques to capture these values and, therefrom, import them into policy decision-making. However, the concern with individual and group constructions could go beyond the redefinition of valued aspects of the environment. The question could be reframed in terms of how actors construct 'good', 'moral', 'ethical' planning. Indeed, it may be a more appropriate use of deliberative techniques to discuss these more fundamental issues, rather than more contingent ones surrounding particular planning decisions.

This also takes seriously the idea of the reflexive individual, developed in professional contexts by Schön and Rein (1994) but generalized in analyses of late modernity (Lash, Szerszynski, and Wynne 1996). Actors are seen as capable of reflecting on these fundamental issues in a meaningful way, so that they can become the arbiters of what counts as ethical action in the policy domain. An institutional discourse approach, therefore, supports the handing of the most profound issues around environmental planning back to actors, returning normative theorizing to the democratic realm. This may prove to be the best hope for deliberative democracy, the discussion of ethical issues concerning the environment at the most fundamental level. This is a precious task for which resources need to be carefully harvested, a realm where communicative rationality can really hold sway provided its potential has not been dissipated in less significant arenas.

APPENDIX

A Rhetorical Approach

Analysing the role of discourse in environmental policy requires a method. Such methods cover both the collection of relevant observations and consistent ways of interrogating the data so collected. Across the discourse literature there is a great variety of methods used. The main distinction is between methods that use some type of coding, often with computer-assisted analysis, and those that rely on close reading to discern trends and patterns. Here the reliance is on close reading, for a number of reasons.

First, coding and computer-assisted analysis is tremendously time-consuming and, of necessity, limits the amount of analysis that can be undertaken. It might be argued that a small amount of rigorous analysis is preferable to a greater quantity of less soundly based analysis. This is a false argument. For use of the computer to organize, aggregate, and summarize coded data does not remove the need for coding, which has to be based on reading by human researchers. It does not, therefore, remove the human element and the subjective decision about what to code and how. Some researchers using repeated codings by independent coders to try to ensure some consistency if not objectivity. But the key decision is not *how* a particular use of language should be coded but *what* should be selected for coding in the first place.

It is, in any case, misleading to represent the identification and coding of a limited set of discursive features as more revealing than the close reading of a complete text. A discourse consists of more than a few isolated grammatical or other linguistic features. Language works by the interconnected parts adding up to more than the sum of those parts. As anyone using quotes knows, context is vital for understanding the text. The human sciences have long known that close reading of the whole text is a better way of understanding that text; only if one is interested in a particular aspect, say the grammatical forms, will coding reveal more than individual readings.

In practice, through close reading, the analyst is using the linguistic capabilities of the brain. The lived experience of the analyst in using language becomes a research skill. What the analyst should do is become more aware than usual of how language is used and interpreted. If the 'normal' understanding of a policy discourse is at issue, then 'normal' but more reflective reading of that discourse is the appropriate research technique. What the social scientist can add to this human scientific method is an understanding of the social context for creating and understanding these discourses, for no text has an inherent meaning completely separate from its context.

However, a commitment to close reading as a method need not mean a 'free-for-all'. There are concepts that can be selectively used to help organize the close reading and that can, therefore, bring order to the categorization of discourses. Within Western thought, rhetoric has long provided such a set of concepts, and a range of analysts have more recently found them useful in understanding a wide variety of discourses (McClosky 1994; Myerson and Rydin 1996*a*; Cantrill and Oravec 1996; Hood 1998). Rhetorical concepts are based on a view of language as essentially bound up in argument and persuasion. As such there are conceptual links to the work of Habermas, who sees persuasive argumentation at the heart of communicative rationality. For rhetoricians, this is as true of written documents and soliloquies as of more obviously rhetorical occasions, such as speeches in a law court. As Hood (1998: 189) argues in his study of public management: 'Rhetoric is central to the way each world-view is sustained, defined and developed as part of a process of challenging other world-views.' Similarly Majone (1989: 7) uses rhetorical analysis because the policy process is above all a process of argumentation.

Within a rhetorical analysis, the emphasis is neither solely on the intentions of the originator of the text, speech, etc., nor on the way in which the audience interprets it, but on the interaction between the two. It is assumed that communication is not a transparent exchange but rather involves difficult work. This is because of barriers to better interpersonal communication at the social, institutional, and individual levels. Therefore, the originator has to try hard to convey the message, uses the resources of her individual creativity with language, the possibilities of the specific arena, and the potential offered by available discourses. Rhetoric helps identify this process at work. It highlights the effort and ease of discursive processes, their interconnectedness, and their development over time with learning, adaptation, and innovation.

The key dimensions and tropes of rhetoric identify certain devices that can be used to persuade the listener. These are:

- Ethos, the personification of credibility of the speaker.
- Pathos, the creation of 'mood music'.
- Logos, the path of argumentation, using the various tropes such as:
  - metaphor (describing one thing in terms of another): 'silky hair', 'policy cascade';
  - synecdoche (taking a part for the whole): one school for all education;
  - metonymy (one thing standing for another): the American flag;
  - irony (literally saying one thing and clearly meaning the opposite): 'of course, I want to go' (the appropriate tone needs to be supplied!)

Where appropriate, these rhetorical devices will be used to clarify the analysis of policy discourse in this book. The aim is not to add to the stamp-collecting tendency of some rhetorical studies (noted by Hood 1998) but to amplify that analysis of policy discourse. Therefore the appropriateness of the rhetorical analysis should be judged by maintaining the balance between understanding the minutiae of the discourse and understanding the overall role that the discourse is playing within the environmental policy process.

On a presentational note, rhetorical analysis can be a very useful analytic tool but unless the text itself is presented, it can often be difficult to replicate or even follow the analysis. Quantitative forms of analysis do not, of course, present their raw data either but they have developed ways of presenting the data in simplified form, a form based on the analysis itself such as standard statistics, graphs, and charts. Here the rhetorical analysis will be accompanied by a similar attempt at simplification through figures identifying the rhetorical tropes at work, examples of key quotations from selected texts, and rhetoric lines tracing the argument through a speech or document, analogous to time-lines in chronological analysis.

REFERENCES

ADAMS, J. (1995), *Risk* (UCL Press: London).
ALM, L. R. (1994), 'Acid Rain and the Key Factors of Issue Maintenance', *Environmental Professional* 16/3: 254–61.
ARMSTRONG, J. (1985), *The Sizewell Report: A New Approach for Major Public Inquiries* (Town and Country Planning Association: London).
AYGEMAN, J., and B. EVANS (eds.) (1994), *Local Environmental Policies and Strategies* (Longman: Harlow, Essex).
BACHRACH, P., and M. S. BARATZ (1962), 'Two Faces of Power', *American Political Science Review* 56: 947–52.
BACOW, L. S., and M. WHEELER (1994), *Environmental Dispute Resolution* (Plenum Press: New York).
BALL, M. (1984), *Housing Policy and Economic Power* (Methuen: London).
BANC (British Association of Nature Conservationists) (1989), *The Conservationists' Response to the Pearce Report* (BANC: London).
BARRETT, S., and C. FUDGE (eds.) (1981), *Policy and Action: Essays on the Implementation of Public Policy* (Methuen: London).
BDP (Building Development Partnership) (1994), *Chester—the Future of an Historic City*, Report for Cheshire County Council, Chester City Council and English Heritage (BDP with MVA Consultants and Donaldsons: London).
BECK, U. (1992), *Risk Society: Towards a New Modernity* (Sage: London).
BELL, M. (1998), *Journal of Environmental Planning and Management* 41/2: 237–51.
BELSTEN, L. A. (1996), 'Environmental Risk Communication and Community Collaboration', in S. A. Muir and T. L. Veenendall (eds.), *Earthtalk: Communication Empowerment for Environmental Action* (Praeger: Westport, Conn.), 27–42.
BERKES, F. (ed.) (1989), *Common Property Resources: An Ecology of Community-Based Sustainable Development* (Belhaven: London).
BEVIR, M. (1999), 'Foucault, Power and Institutions', *Political Studies* 47: 345–59.
BHATTI, M. (ed.) (1994), *Housing and the Environment: A New Agenda*, Institute of Housing: London.
BICKERSTAFF, K., and G. WALKER (1999), 'Cleaning the Smog? Public Responses to Air Quality Information', *Local Environment* 4/3: 279–94.
——(forthcoming), 'Risk, Responsibility and Blame: Analysing Vocabularies of Motive in Air Pollution(ing) Discourses', *Environment and Planning A*.
BOHMAN, J. (1996), *Public Deliberation: Pluralism, Cmplexity and Democracy* (MIT Press: Cambridge, Mass.).
BRAMLEY, I. (1985), 'Style in Planning Documents—An Approach from Linguistics', *The Planner*, August, 8–10.
BRIDGER, J. C. (1996), 'Community Imagery and the Built Environment', *The Sociological Quarterly* 37/3: 353–74.
BRYSON, J., and B. CROSBY (1993), 'Policy Planning and the Design and Use of Forums, Arenas and Courts', *Environment and Planning B* 20: 175–94.
BURGESS, J., C. HARRISON, and P. FILIUS (1998), 'Environmental Communication and the Cultural Politics of Environmental Citizenship', *Environment and Planning A* 30: 1445–60.

Burgess, J., C. Harrison, and P. Maiteny (1991), 'Contested Meanings: The Consumption of News about Nature Conservation', *Media, Culture and Society* 13: 499–519.

Burke, K. (1969), *A Rhetoric of Motives* (University of Los Angeles Press, Los Angeles).

Burningham, K. (2000), 'Using the Language of NIMBY: A Topic for Research, not an Activity for Researchers', *Local Environment* 5/1: 55–68.

CAG Consultants and Land Use Consultants (1997), 'Environmental Capital: A New Approach', Report prepared for Countryside Commission, English Heritage, English Nature and Environment Agency.

——(2001), 'Quality of Life Capital: Managing Environmental, Social and Economic Benefits. Overview Report', prepared for Countryside Agency, English Heritage, English Nature, and Environment Agency.

Cantrill, J. G. (1996), 'Gold, Yellowstone and the Search for a Rhetorical Identity', in C. Herndl and S. Brown (eds.), *Green Culture: Environmental Rhetoric in Contemporary America* (University of Wisconsin Press: Madison), 166–94.

Cantrill, J. G., and C. L. Oravec (eds.) (1996), *The Symbolic Earth: Discourse and our Creation of the Environment* (University Press of Kentucky: Lexington).

Chapman, G., K. Kumar, C. Fraser, and I. Gaber (1997), *Environmentalism and the Mass Media: The North–South divide* (Routledge: London).

Chong, D. (1991), *Collective Action and the Civil Rights Movement* (University of Chicago Press: Chicago).

Coanus, T., F. Duchene, and E. Martinais (1998), 'Les Relations des gestionnaires du risque urbain avec les populations riveraines. Critique d'une certain idée de la communication'. Paper to the Annual Risk Analysis Conference, Paris, 12–14 October.

Coleman, J. S. (1988), 'Social Capital in the Creation of Human Capital', *American Journal of Sociology* 94 (suppl.), 95–119.

Connell, B. (1999), 'Accommodating Development: A View from West Sussex; Environmental Capacity and Strategic Development', Paper to ESRC Planning, Space, and Sustainability Seminar II: Concepts and Tools, Cardiff University.

Coyle, D. J. (1997), 'A Cultural Theory of Organizations', in R. J. Ellis and M. Thompson (eds.), *Culture Matters: Essays in Honour of Aaron Wildavsky* (Westview Press: Boulder, Col.), 59–78.

Crenson, M. A. (1971), *The Unpolitics of Air Pollution: A Study of Non-Decisionmaking in the Cities* (Johns Hopkins University Press: Baltimore).

Crosby, N. (1995), 'Citizens Juries: One Solution for Difficult Environmental Problems', in O. Renn, T. Webler, and P. Wiedemann (eds.), *Fairness and Competence in Citizen Participation* (Kluwer: Dordrecht).

Crowfoot, J., and J. Wondolleck (1990), *Environmental Disputes* (Island Press: Washington, DC).

Dalby, S., and F. Mackenzie (1997), 'Reconceptualising Local Community: Environment, Identity and Threat', *Area* 29/2: 99–108.

Daly, H. (1977), *Steady-State Economics* (W. H. Freeman: San Francisco).

Darier, E. (ed.) (1999), *Discourses of the Environment* (Basil Blackwell: Oxford).

Darlow, A., and L. Newby (1997), 'Partnerships: Panacea or Pitfall? Experience in Leicester Environment City', *Local Environment* 2/1: 73–82.

DEFRA (Department of Environment, Food and Rural Affairs) (2001), *Air Pollution—What it Means for Your Health* (DEFRA: London).

De La Court, T. (1990), *Beyond Brundtland: Green Development in the 1990s* (Zed Books: London).

de Leon, P. (1999), 'The Stages Approach to the Policy Process: What Has It Done? Where Is It Going?', in P. Sabatier (ed.), *Theories of the Policy Process* (Westview Press: Boulder, Col.).

De Luca, K. M. (1999), *Image Politics: The New Rhetoric of Environmental Activism* (The Guildford Press: New York).

DETR (Department of Environment, Transport and the Regions) (1997), *The Application of Environmental Capacity to Land Use Planning*, Report by ENTEC UK Ltd. (HMSO: London).

——(2000) *Integrated Pollution Prevention and Control: A Practical Guide*, 1st edn. (DETR: London).

Dobson, A. (1998), *Justice and the Environment: Conceptions of Environmental Sustainability and Dimensions of Social Justice* (Oxford University Press: Oxford).

DoE (Department of the Environment) (1991), *Policy Appraisal and the Environment* (HMSO: London).

——(1993), *Making Markets Work for the Environment* (HMSO: London).

Douglas, M. (1966), *Purity and Danger: An Analysis of the Concepts of Pollutions and Taboo* (Routledge & Kegan Paul: London).

Douglas, M., and A. Wildavsky (1983), *Risk and Culture: An Essay on the Selection of Technological and Environmental Dangers* (University of California Press: Berkeley).

Douglas, S. (1992), 'Negotiation Audiences: The Role of the Mass Media', in L. Putnam and M. Roloff (eds.), *Communication and Negotiation* (Sage: London), 250–72.

Dowding, K. (1991), *Rational Choice and Political Power* (Edward Elgar: Aldershot).

Dowding, K., and P. Dunleavy (1996), 'Production, Disbursement and Consumption: The Modes and Modalities of Goods and Services', in S. Edgell, K. Hetherington, and S. Warde (eds.), *Consumption Matters* (Basil Blackwell: Oxford), 36–65.

Dryzek, J. (1990), *Discursive Democracy: Politics, Policy and Political Science* (Cambridge University Press: Cambridge).

——(1995), Political and Ecological Communication', *Environmental Politics* 4/4: 13–30.

——(1997), *The Politics of the Earth: Environmental Discourses* (Oxford University Press: Oxford).

——(2000), *Deliberative Democracy and Beyond: Liberals, Critics, Contestations* (Oxford University Press: Oxford).

Duane, T. (1997), 'Community Participation in Ecosystem Management', *Ecology Law Quarterly* 24: 771–97.

Dunleavy, P. (1996), *Democracy, Bureaucracy and Public Choice* (Harvester-Wheatsheaf: Hemel Hempstead).

Dwyer, C., and I. Hodge (1996), *The Countryside in Trust* (Harvester/Wheatsheaf: London).

EA (Environment Agency) (1998), *Consensus Building for Sustainable Development* (Environment Agency: Bristol).

Edelman, M. (1984), 'The Political Language of the Helping Professions', in M. Shapiro (ed.), *Language and Politics* (Basil Blackwell: Oxford), 44–60.

——(1988), *Constructing the Political Spectacle* (University of Chicago Press: Chicago).

Evans, B., and Y. Rydin (1997), 'Planning, Professionalism and Sustainability', in A. Blowers and B. Evans (eds.), *Town Planning into the 21st Century* (Routledge: London), 55–70.

Farrell, T. (1993), *Norms of Rhetorical Culture* (Yale University Press: New Haven).

Fischer, F. (2000), *Citizens, Experts and the Environment: The Politics of Local Knowledge* (Duke University Press: Durham).

Flyvbjerg, B. (1998*a*), *Rationality and Power: Democracy in Practice* (University of Chicago Press: Chicago).

——(1998b), 'Habermas and Foucault: Thinkers for Civil Society?', *British Journal of Sociology* 49/2: 210–32.

Forester, J. (1989), *Planning in the Face of Power* (University of California Press: Berkeley).

Foster, J. (ed.) (1997), *Valuing Nature? Economics, Ethics and Environment* (Routledge: London).

Foucault, M. (1978), *The History of Sexuality*, i. *An Introduction*, transl. R. Hurley (Pantheon: New York).

——(1984), The Order of Discourse', in M. Shapiro (ed.), *Language and Politics* (Basil Blackwell: Oxford), 108–38.

Galbraith, K. (1952), *American Capitalism* (Hamilton: London).

Gibbons, P., J. Bradac, and J. Busch (1992), 'The Role of Language in Negotiations: Threats and Promises', in L. Putnam and M. Roloff (eds.), *Communication and Negotiation* (Sage: London), 156–75.

Giddens, A. (1990), *The Consequences of Modernity* (Polity: Cambridge).

Gilbert, R., D. Stevenson, H. Girardet, and R. Stren (1996), *Making Cities Work: The Role of Local Authorities in the Urban Environment* (Earthscan: London).

GLA (Greater London Authority) (2001) *Clean Air for London: Highlights of the Mayor's Draft Air Quality Strategy* (GLA: London).

Glasbergen, P. (ed.) (1995), *Managing Environmental Disputes: Network Management as an Alternative* (Kluwer Academic: Dordrecht).

Gouldson, A., and J. Murphy (1998), *Regulatory Realities: The Implementation and Impact of Industrial Environmental Regulation* (Earthscan: London).

Grant, J. (1994), 'On Some Public Uses of Planning "Theory": Rhetoric and Expertise in Community Planning Disputes', *Town Planning Review* 65/1: 59–76.

Green, J. (1999), *Deep Democracy: Community, Diversity, and Transformation* (Rowman & Littlefield: Lanham, Md.).

Green, D., and I. Shapiro (1994), *Pathologies of Rational Choice Theory* (Yale University Press: New Haven).

Grigson, S. (1995), *The Limits of Environmental Capacity* (Barton Willmore Partnerships & HBF: London).

Habermas, J. (1984), *The Theory of Communicative Action*, transl. T. McCarthy (Heinemann: London), i.

——(1987), *The Theory of Communicative Action*, transl. T. McCarthy (Polity: Cambridge), ii.

——*Between Facts and Norms* (Polity: Oxford).

Hajer, M. (1995), *The Politics of Environmental Discourse* (Oxford University Press: Oxford).

Hannigan, J. (1995), *Environmental Sociology: A Social Constructionist Perspective* (Routledge: London).

HANSEN, A. (1991), 'The Media and the Social Construction of the Environment', *Media Culture and Society* 13/4: 443–58.

——(ed.) (1993), *The Mass Media and Environmental Issues* (Leicester University Press: Leicester).

HARDOY, J., P. MILTON, and D. SATTERTHWAITE (1992), *Environmental Problems in Third World Cities* (Earthscan: London).

HARRISON, C., and J. BURGESS (1994), 'Social Constructions of Nature: A Case Study of the Conflicts Over Rainham Marshes SSSI', *Transactions of the Institute of British Geographers* 19: 291–310.

HARVEY, D. (1989), *The Condition of Postmodernity* (Basil Blackwell: Oxford).

——(1996), *Justice, Nature and the Geography of Difference* (Basil Blackwell: Oxford).

HEALEY, P. (1993), 'The Communicative Work of Development Plans', *Environment and Planning B* 20: 83–104.

——(1996), 'The Communicative Turn in Planning Theory and its Implications for Spatial Strategy Formation', *Environment and Planning B* 23: 217–34.

——(1997), *Collaborative Planning: Shaping Places in Fragmented Societies* (Macmillan: London).

HEALEY, P., and J. HILLIER (1996), 'Communicative Micropolitics: A Story of Claims and Discourses', *International Planning Studies* 1/2: 165–84.

HEALEY, P., P. MCNAMARA, M. ELSON, and J. DOAK (1988), *Land Use Planning and the Mediation of Urban Change* (Cambridge University Press: Cambridge).

HEMMATI, M. (2001), *Multi-stakeholder Processed for Governance and Sustainability: Beyond Deadlock and Conflict* (Earthscan: London).

Her Majesty's Government (1999), *A Better Quality of Life*, Cm. 4345 (The Stationery Office: London).

HERNDL, C. G., and S. C. BROWN (1996), *Green Culture: Environmental Rhetoric in Contemporary America* (University of Wisconsin Press: Madison).

HILLIER, J. (1993), 'To Boldly Go Where No Planners Have Ever . . .' *Environment and Planning D* 11: 89–113.

——(1998), 'Beyond Confused Noise: Ideas Towards Communicative Procedural Justice', *Journal of Planning Education and Research* 18: 14–24.

HILLIER, J., and T. VAN LOOIJ (1997), 'Who Speaks for the Poor?', *International Planning Studies* 2/1: 7–25.

HOLZINGER, K. (2001), 'Limits of Co-operation: A German Case of Environmental Mediation', *European Environment*.

HOOD, C. (1998), *The Art of the State* (Oxford University Press: Oxford).

INCE, M. (1984), *Sizewell Report: What Happened at the Inquiry?* (Pluto: London).

INGHAM, Z. (1996), 'Landscape, Drama, and Dissensus: The Rhetorical Education of Red Lodge, Montana', in C. Herndl and S. Brown (eds.), *Green Culture: Environmental Rhetoric in Contemporary America* (University of Wisconsin Press: Madison), 195–212.

INGOLD, T. (1993), 'Globes and Spheres: The Topology of Environmentalism', in K. Milton (ed.), *Environmentalism: The View from Anthrolopology* (Routledge: London), 31–42.

INNES, J., and D. BOOHER (1996), 'Consensus-Building as Role-Playing and Bricolage—Towards a Theory of Collaborative Planning', Paper to the ACSP/AESoP Congress, Toronto, Canada.

IRWIN, A. (1995), *Citizen Science: A Study of People, Expertise and Sustainable Development* (Routledge: London).

JACOBS, M. (1997*a*), 'Environmental Valuation, Deliberative Democracy and Public Decision-Making Institutions', in J. Foster (ed.), *Valuing Nature?* (Routledge: London).

——(1997*b*), *Making Sense of Environmental Capacity* (Council for the Protection of Rural England: London).

JÄNICKE, M. (1996), 'The Political System's Capacity for Environmental Policy', in M. Jänicke and H. Weidner (eds.), *National Environmental Policies: A Comparative Study of Capacity-Building* (Springer: Helsinki).

JOHNSON, T. (1972), *Professions and Power* (Macmillan: Basingstoke).

KATZ, S., and C. MILLER (1996), 'The Low-Level Radioactive Waste Siting Controversy in North Carolina: Toward a Rhetorical Model of Risk Communication', in C. Herndl and S. Brown (eds.), *Green Culture: Environmental Rhetoric in Contemporary America* (University of Wisconsin Press: Madison), 111–40.

KECK, M. E., and K. SIKKINK (1999), 'Transnational Advocacy Networks in International and Regional Politics', *International Social Science Journal* 159: 89–101.

KEMP, R. (1985), 'Planning, Public Hearings and the Politics of Discourse', in J. Forester (ed.), *Critical Theory and Public Life* (MIT Press: Cambridge Mass.), 177–201.

KEOUGH, C. (1992), 'Bargaining Arguments and Argumentative Bargainers', in L. Putnam and M. Roloff (eds.), *Communication and Negotiation* (Sage: London), 109–27.

KEYNES, J. M. (1936), *The General Theory of Employment, Interest and Money*, repr. 1973 (Macmillan: London).

KILLINGSWORTH, M. J., and J. S. PALMER (1992), *Ecospeak: Rhetoric and Environmental Politics in America* (Southern Illinois University Press: Carbondale).

——(1996), 'Millenial Ecology: The Apocalyptic Narrative from *Silent Spring* to *Global Warming*', in C. Herndl and S. Brown (eds.), *Green Culture: Environmental Rhetoric in Ccontemporary America* (University of Wisconsin Press: Madison), 21–45.

KINGDON, J. W. (1984), *Agendas, Alternatives and Public Policies* (Little Brown & Co.: Boston).

LACEY, C., and D. LONGMAN (1993), 'The Press and Public Access to the Environment and Development Debate', *The Sociological Review* 41/2: 207–43.

LAFFERTY, W. (ed.) (2001), *Sustainable Communities in Europe* (Earthscan: London).

LAFFERTY, W., and K. ECKERBERG (eds.) (1998), *From Earth Summit to Local Agenda 21: Working Towards Sustainable Development* (Earthscan: London).

LARSEN, H. (1997), *Foreign Policy and Discourse Analysis France, Britain and Europe* (Routledge: London).

LASALA (2001), *Accelerating Local Sustainability: Evaluation European Local Agenda 21 Processes.* Report of the LASALA (Local Authorities' Self-Assessment of Local Agenda 21) project (ICLEI: Freiburg).

LASH, S., B. SZERSZNYSKI, and B. WYNNE (1996), *Risk, Environment and Modernity: Towards a New Ecology* (Sage: London).

LATOUR, B. (1999), *Pandora's Hope: Essays on the Reality of Science Studies* (Harvard University Press: Cambridge, Mass.).

LAURIA, M., and M. J. SOLL (1996), 'Communicative Action, Power and Misinformation in a Site Selection Process', *Journal of Planning Education and Research* 15/3: 199–211.
LEVY, N. (1999), 'Foucault's Unnatural Ecology', in E. Darier (ed.), *Discourse of the Environment* (Basil Blackwell: Oxford), 203–16.
LINDBLOM, C. E. (1977), *Politics and Markets* (Basic Books: New York).
LIPSKY, M. (1980), *Street-Level Bureaucracy* (Russell Sage Foundation: New York).
LOWNDES, V. (1999), 'Management Change in Local Governance', in G. Stoker (ed.), *The New Management of British Local Governance* (Macmillan: Basingstoke), 22–39.
LUKE, T. W. (1999), 'Environmentality as Green Governmentality' in E. Darier (ed.), *Discourses of the Environment* (Basil Blackwell: Oxford), 121–51.
MCCLOSKY, D. (1994), *Knowledge and Persuasion in Economics* CUP: Cambridge.
MACDONALD, K. (1995), *The Sociology of Professions* (Sage: London).
MCLAREN, D., S. BULLOCK, and N. YOUSUF (1998), *Tomorrow's World: Britain's Share in a Sustainable Future* (Earthscan: London).
MACNAGHTEN, P., and J. URRY (1998), *Contested Natures* (Sage: London).
MAJONE, G. (1989), *Evidence, Argument and Persuasion in the Policy Process* (Yale University Press: New Haven).
MARCH, J., and J. OLSEN (1989), *Rediscovering Institutions: The Organizational Basis of Politics* (The Free Press: New York).
MARTINEZ-ALLIER, J. (1987), *Ecological Economics: Energy, Environment and Society* (Basil Blackwell: Oxford).
MASON, M. (1999), *Environmental Democracy* (Earthscan: London).
MATTHEWS, D. R. (1996), 'Mere Anarchy? Canada's "Turbot War" as the Moral Regulation of Nature', *Canadian Journal of Sociology* 21/4: 505–22.
MAZZA, L. (1986), 'Guistificazione e autonomia degli elementi di piano', *Urbanistica* 82: 56–63.
MEADOWS, D., D. MEADOWS, J. RANDERS, and W. BEHRENS (1972), *The Limits to Growth* (Universe Books: New York).
MILLER, M., and B. PARNELL REICHERT (2000), 'Interest Group Strategies and Journalistic Norms', in S. Allan, B. Adam, and C. Carter (eds.), *Environmental Risks and the Media* (Routledge: London), 45–54.
MILTON, K. (ed.) (1993), *Environmentalism: The View from Anthropology* (Routledge: London).
——(1996), *Environmentalism and Cultural Theory: Exploring the Role of Anthropology in Environmental Discourse* (Routledge: London).
MIRANDA, L., and M. HORDIJK (1998), 'Let Us Build Cities for Life: The National Campaign of Local Agenda 21s in Peru', *Environment and Urbanisation* 10/2: 69–102.
MITTLER, D. (1999), 'Environmental Space and Barriers to Local Sustainability: Evidence from Edinburgh, Scotland', *Local Environment* 4/3: 353–65.
MUIR, S. A., and T. L. VEENENDALL (1996), *Earthtalk: Communication Empowerment for Environmental Action* (Praeger: Westport, Conn.).
MURDOCH, J., S. ABRAMS, and T. MARSDEN (1999), 'Modalities of Planning: A Reflection on the Persuasive Powers of the Development Plan', *Town Planning Review* 70/2: 191–212.
——(2000), 'Technical Expertise and Public Participation in Planning for Housing: "Playing the Numbers Game"' in G. Stoker (ed.), *The New Politics of British Local Governance* (Macmillan: Basingstoke), 198–214.

MYERSON, G., and Y. RYDIN (1994), '"Environment" and Planning: A Tale of the Mundane and the Sublime', *Environment and Planning D* 12: 437–52.

——(1996*a*), *The Language of Environment: A New Rhetoric* (UCL Press: London).

——(1996*b*), 'Sustainable Development: The Implications of the Global Debate for Land Use Planning', in S. Buckingham-Hatfield and B. Evans (eds.), *Environmental Planning and Sustainability* (Wiley & Sons: Chichester), 19–34.

National Audit Office (2001), *Policy Development: Improving Air Quality* (The Stationery Office: London).

NEALE, A. (1997), 'Organizing Environmental Self-Regulation: Liberal Governmentality and the Pursuit of Ecological Modernisation in Europe', *Environmental Politics* 6/4: 1–24.

OLSON, M. (1969), *The Logic of Collective Action* (Harvard University Press: Cambridge, Mass.).

O'NEILL, J. (1993), *Ecology, Policy and Politics: Human Well-being and the Natural World* (Routledge: London).

OPIE, J., and N. ELLIOT (1996), 'Tracking the Elusive Jeremiad: The Rhetorical Character of American Environmental Discourse', in J. G. Cantrill and C. L. Oravec (eds.), *The Symbolic Earth: Discourse and Our Creation of the Environment* (University Press of Kentucky: Lexington), 9–37.

O'RIORDAN, T., R. KEMP, and M. PURDUE (1988), *Sizewell B* (Macmillan: Basingstoke).

O'RIORDAN, T., and J. CAMERON (eds.) (1994), *Interpreting the Precautionary Principle* (Earthscan: London).

OSTROM, E. (1990), *Governing the Commons: The Evolution of Institutions for Collection Action* (Cambridge University Press: Cambridge).

——(1992), *Crafting Institutions for Self-governing Irrigation Systems* (ICS Press: San Francisco).

——(1995), 'Constituting Social Capital and Collective Action', in R. O. Keohane and E. Ostrom (eds.), *Local Commons and Global Interdependence* (Sage: London), 125–60.

——(1999), 'Institutional Rational Choice: An Assessment of the Institutional Analysis and Development Framework', in P. Sabatier (ed.), *Theories of the Policy Process* (Westview Press: Boulder, Col.), 35–72.

OSTROM, E., R. GARDNER, and J. WALKER (1994), *Rules, Games and Common-Pool Resources* (University of Michigan Press: Ann Arbor).

OSTROM, V., D. FEENY, and H. PICHT (1988), 'Institutional Analysis and Development: Rethinking the Terms of Choice', in V. Ostrom, D. Feeny, and H. Picht (eds.), *Rethinking Institutional Analysis and Development* (International Centre for Economic Growth: San Francisco), 439–66.

OWENS, S. (1994), 'Land, Limits and Sustainability: A Conceptual Framework and Some Dilemmas for the Planning System', *Transaction of the Institute of British Geographers* 19/4: 439–56.

Pastille Project (2001), *Promoting Action for Sustainability Through Indicators at the Local Level in Europe: Interim Report*, Report to European Commission (LSE: London).

PATTERSON, A., and K. THEOBALD (1995), 'Sustainable Development, Agenda 21 and the New Local Governance in Britain', *Regional Studies* 29/8: 773–8.

PENNINGTON, M. (1997), 'Budgets, Bureaucrats and the Containment of Urban England', *Environmental Politics* 6/4: 76–107.

PERELMAN, C. (1982), *The Realm of Rhetoric* (University of Notre Dame Press: Notre Dame, Ind.).

PHELPS, N., and M. TEWDWR-JONES (1999), 'Discourse and Distortion in Collaborative and Associative Governance', Paper to Discourse and Policy Change Conference, Glasgow University.

——(2000), 'Scratching the Surface of Collaborative and Associative Governance: Identifying the Diversity of Social Action in Institutional Capacity Building', *Environment and Planning A* 32/1: 111–30.

POCOCK, J. (1984), 'Verbalizing a Political Act: Toward a Politics of Speech', in M. Shapiro (ed.), *Language and Politics* (Basil Blackwell: Oxford), 25–43.

POOLE, M. S., D. SHANNON, and G. DESANCTIS (1992), 'Communication Media and Negotiation Processes', in L. Putnam and M. Roloff (eds.), *Communication and Negotiation* (Sage: London).

POWER, M. (1997), *The Audit Society: Rituals of Verification* (Clarendon Press: Oxford).

PUTNAM, L., and M. E. ROLOFF (1992), *Communication and Negotiation* (Sage: London).

REDCLIFT, M., and T. BENTON (eds.) (1994), *Social Theory and the Global Environment* (Routledge: London).

RICHARDS, P. (1993), 'Natural Symbols and Natural History: Chimpanzees, Elephants and Experiments in Mende Thought', in K. Milton (ed.), *Environmentalism: The View from Anthropology* (Routledge: London), 144–59.

RICHARDSON, J. (1994), 'EU Water Policy: Uncertain Agendas, Shifting Networks and Complex Coalitions', *Environmental Politics* 3/4: 139–67.

RICHARDSON, T. (1996), 'Foucauldian Discourse: Power and Truth in Urban and Regional Policy Making', *European Planning Studies* 4/3: 279–92.

——(1999), 'Constructing Rationalities for Sustainable Planning: Power and Strategic Environmental Assessment', Paper to ESRC Planning, Space and Sustainability Seminar II: Concepts and Tools, Cardiff University.

RIKER, W. (1986), *The Art of Political Manipulation* (Yale University Press: New Haven).

——(ed.) (1993), *Agenda Formulation* (University of Michigan Press: Ann Arbor).

ROBERTS, I. (2000), 'Leicester Environment City: Learning How to Make Local Agenda 21, Partnerships and Participation Deliver', *Environment and Urbanization* 12/2: 9–26.

RYDIN, Y. (1986), *Housing Land Policy* (Gower: Aldershot).

——(1998*a*), 'Land Use Planning and Environmental Capacity: Reassessing the Use of Regulatory Policy Tools to Achieve Sustainable Development', *Journal of Environmental Planning and Management* 41/6: 749–65.

——(1998*b*), '"Managing Urban Air Quality": Language and Rational Choice in Metropolitan Governance', *Environmental and Planning A* 30: 1429–43.

——(1998*c*), *Urban and Environmental Planning in the UK* (Macmillan: Basingstoke).

——(2000), *The Public and Local Environmental Policy: Strategies for Promoting Public Participation*, Town and Country Planning Association Tomorrow 2000 Series (TCPA: London).

RYDIN, Y., and G. MYERSON (1989), 'Explaining and Interpreting Ideological Effects: A Rhetorical Approach to Green Belts', *Environment and Planning D* 7: 463–79.

RYDIN, Y., and PENNINGTON, M. (2000), 'Public Participation and Local Environmental

Planning: The Collective Action Problem and the Potential of Social Capital', *Local Environment* 5/2: 153–69.

——(2001), 'Discourse of the Prisoner's Dilemma: The Role of the Local Press in Environmental Policy', *Environmental Politics*.

RYDIN, Y., and F. SOMMER (1999), *Evaluation of Stages 1 & 2 of the LITMUS Project: An Evaluation of the Public Participation Dimensions of a Local Sustainability Indicators Project*, Report to London Borough of Southwark (LSE: London).

SABATIER, P. (ed.) (1999), *Theories of the Policy Process* (Westview Press: Boulder, Col.).

SABATIER, P., and H. C. JENKINS-SMITH (1993), *Policy Change and Learning: An Advocacy Coalition Approach* (Westview Press: Boulder, Col.).

——(1999), 'The Advocacy Coalition Framework: An Assessment', in P. Sabatier (ed.), *Theories of the Policy Process* (Westview Press: Boulder, Col.), 117–33.

SACHSTMAN, D. (1996), 'The Mass Media "Discover" the Environment: Influences on Environmental Reporting in the First Twenty Years', in J. G. Cantrill and C. L. Oravec (eds.), *The Symbolic Earth: Discourse and Our Creation of the Environment* (University Press of Kentucky: Lexington), 241–56.

SCHATTSCHNEIDER, E. (1960), *The Semi-Sovereign People* (Holt, Rinehart, & Winston: New York).

SCHLECHTWEG, H. (1996), 'Media Frames and Environmental Discourse: The Case of "Focus Logjam"' in J. G. Cantrill and C. L. Oravec (eds.), *The Symbolic Earth: Discourse and Our Creation of the Environment* (University Press of Kentucky: Lexington), 257–77.

SCHÖN, D., and M. REIN (1994), *Frame Reflection: Towards the Resolution of Intractable Policy Controversies* (Basic Books: New York).

SCHUMACHER, E. (1974), *Small is Beautiful: A Study of Economics as if People Mattered* (Sphere Books: London).

SEIDMAN, S. (ed.) (1994), *The Postmodern Turn: new perspectives on social theory* (CUP: Cambridge).

SELMAN, P. (1998), 'Local Agenda 21: Substance or Spin?' *Journal of Environmental Planning and Management* 41/5: 533–53.

SELMAN, P., and J. PARKER (1997), 'Citizenship, Civicness and Social Capital Local Agenda 21', *Local Environment* 2/2: 171–84.

——(1999), Tales of Local Sustainability', *Local Environment* 4/1: 47–60.

SENECAH, S. (1996), 'Forever Wild or Forever in Battle: Metaphors of Empowerment in the Continuing Controversy over the Adirondacks', in S. Muir and T. Veenendall (eds.), *Earthtalk: Communication Empowerment for Environmental Action* (Praeger: Westport, Conn.), 95–118.

SHARP, L. (1999), 'Local Policy for the Global Environment: In Search of a New Perspective', *Environmental Politics* 8/4: 137–59.

SHERIDAN, A. (1980), *Foucault: The Will to Truth* (Tavistock: London).

SHOVE, E., and S. GUY (2000), *The Sociology of Energy, Buildings and the Environment: Constructing Knowledge, Designing Practice* (Routledge: London).

SHUCKSMITH, M. (1990), *Housebuilding in Britain's Countryside* (Routledge: London).

SIMMONS, C., and N. CHAMBERS (1998), 'Footprinting UK Households: How Big is Your Ecological Garden?', *Local Environment* 3/3: 355–62.

SMITH, G., and C. WALES (2000), 'Citizens' Juries and Deliberative Democracy', *Political Studies* 48/1: 51–65.

SOMMER, F. (2000), 'Monitoring and Evaluating Outcomes of Community Involvement—the LITMUS Experience', *Local Environment* 5/4: 483–91.

SPANGLE, M., and D. KNAPP (1996), 'Ways We Talk about the Earth: An Exploration of Persuasive Tactics and Appeals in Environmental Discourse' in S. Muir and T. Veenendall (eds.), *Earthtalk: Communication Empowerment for Environmental Action* (Praeger: Westport, Conn.), 3–26.

STOKER, G. (ed.) (1999), *The New Management of British Local Governance* (Macmillan: Basingstoke).

STONE, C. (1989), *Regime Politics: Governing Atlanta 1946–88* (University of Kansas Press: Lawrence).

STONE, D. (1989), 'Causal Stories and the Formation of Policy Agendas', *Political Science Quarterly* 104/2: 281–300.

SUSSKIND, L., and G. MCMAHON (1985), 'The Theory and Practice of Negotiated Rulemaking', *Yale Journal on Regulation* 3/1: 133–65.

TAYLOR, P., and F. BUTTELL (1992), 'How Do We Know We Have Global Environmental Problems? Science and the Globalisation of Environmental Discourse', *Geoforum* 23/3: 405–16.

TAYLOR, M., and H. WARD (1982), 'Chickens, Whales and Lumpy Public Goods: Alternative Models of Public Goods Provision', *Political Studies* 30: 350–70.

TEWDWR-JONES, M., and P. ALLMENDINGER (1998), 'Deconstructing Communicative Rationality: A Critique of Habermasian Collaborative Planning', *Environment and Planning A* 30: 1975–89.

THROGMORTON, J. (1991), 'The Rhetorics of Policy Analysis', *Policy Sciences* 24: 153–79.

TROJA, M. (2001), 'Capacity Building in Environmental Policy through Mediation—Especially from the Mediation Project "Waste Management Project of Berlin"' *European Environment*.

TUTZAUER, F. (1992), 'The Communication of Offers in Dyadic Bargaining', in L. Putnam and M. Roloff (eds.), *Communication and Negotiation* (Sage: London), 67–82.

VAN DIJK, T. (1996), *Discourse Studies: A Multi-Discipline Introduction* (Sage: London).

VAN WOERKUM, C. (2002), 'Orality in Environmental Planning', *European Environment* 12: 160–72.

VIEHÖVER, W. (2001), 'Political Negotiation and Co-operation in the Shadow of Public Discourse: The Formulation of the German Waste Management System DSD as a Case Study', *European Environment*.

VOGEL, D. (1986), *National Styles of Regulation: Environmental Policy in Great Britain and the United States* (Cornell University Press: Ithaca).

VOISEY, H., C. BEUERMANN, L. ASTRID, and T. O'RIORDAN (1996), 'The Political Significance of Local Agenda 21: The Early Stages of some European Experience', *Local Environment* 1: 33–50.

WACKERNAGEL, M., and W. REES (1996), *Our Ecological Footprint* (New Society Publications: Gabriola Island).

WADDELL, C. (1996), 'Saving the Great Lakes: Public Participation in Environmental Policy', in C. Herndl and S. Brown (eds.), *Green Culture: Environmental Rhetoric in Contemporary America* (University of Wisconsin Press: Madison), 141–65.

WALL, D. (1999), *Earth First! And the Anti-Roads Movement* (Routledge: London).

WCED (World Commission on Environment and Development) (1987), *Our Common Future* (Oxford University Press: Oxford).

WEAVER, B. (1996), '"What to do with the Mountain People?": The Darker Side of the Successful Campaign to Establish the Great Smoky Mountains National Park', in J. G. Cantrill and C. L. Oravec (eds.) (1996), *The Symbolic Earth: Discourse and Our Creation of the Environment* (University Press of Kentucky: Lexington), 151–75.

VON WEISACKER, E., A. LOVINS, and L. LOVINS (1998), *Factor Four: Doubling Wealth, Having Resource Use* (Earthscan: London).

West Sussex County Council (1996), *Environmental Capacity in West Sussex: Pressures on the Countryside, Towns and Quality of Life*, West Sussex County Council Structure Plan 3rd Review, Study prepared by West Sussex County Planning Department, Chester.

WHATMORE, S., and S. BOUCHER (1993), 'Bargaining with Nature: The Discourse and Practice of "Environmental Planning Gain"', *Transactions of the Institute of British Geographers* NS 18: 166–78.

WILD, A., and R. MARSHALL (1999), 'Participatory Practice in the Context of Local Agenda 21: A Case Study Evaluation of Experience in Three English Local Authorities', *Sustainable Development* 7/3: 151–62.

WILDER, R. J. (1993), 'Cooperative Governance, Environmental Policy and the Management of Offshore Oil and Gas in the United States', *Ocean Development and International Law* 24/1: 41–62.

WILSON, J. (1996), 'Metaphors, Growth Coalition Discourses and Black Poverty Neighbourhoods in a US City', *Antipode* 28: 72–96.

WYKES, M. (2000), 'The Burrowers: News about Bodies, Tunnels and Green Guerrillas', in S. Allan, B. Adam, and C. Carter (eds.), *Environmental Risks and the Media* (Routledge: London), 73–89.

WYNNE, G. (1996), 'May the Sheep Safely Graze? A Reflexive View of the Expert–Lay Knowledge Divide', in S. Lash, B. Szersznyski, and B. Wynne, *Risk, Environment and Modernity: Towards a New Ecology* (Sage: London), 44–83.

YANOW, D. (1996), *How Does a Policy Mean? Interpreting Policy and Organizational Action* (Georgetown University Press: Washington, DC).

YEARLEY, S. (1991), *The Green Case: A Sociology of Environmental Issues, Arguments and Politics* (Harper Collins: London).

YOUNG, S. (1996), 'Stepping Stones to Empowerment? Participation in the Context of Local Agenda 21', *Local Government Policy Making* 22: 25–31.

——(1997), 'Community-Based Partnerships and Sustainable Development: A Third Force in the Social Economy', in S. Baker, M. Kousis, D. Richardson, and S. Young (eds.), *The Politics of Sustainable Development* (Routledge: London).

——(2000), 'Participation Strategies and Local Environmental Politics: Local Agenda 21', in G. Stoker (ed.), *The New Politics of British Local Governance* (Macmillan: Basingstoke).

INDEX